AquaGuide

Catfish

Dr Jürgen Schmidt

CONTENTS

First published in the UK in 2003 by Interpet Publishing,
Vincent Lane, Dorking, Surrey RH4 3YX, England

English text © 2003 by Interpet Publishing Ltd.

ISBN 1-84286-084-4

The recommendations in this book are given without any guarantees on the part of the author and publisher. If in doubt, seek the advice of a vet or aquatic specialist.

Translation: Heike de Ste. Croix
Technical consultant: Richard Hardwick
Originally published in Germany in 2000 by bede-Verlag,
Bühlfelderweg 12, D-94239 Ruhmannsfelden
© 2000 by bede-Verlag
Typeset by MATS, Southend-on-Sea, Essex

Picture Credits
The pictures were supplied by Aqualife Taiwan, H. Hieronimus, M.-P. & C. Piednoir, Dr J. Schmidt or bede-Verlag, as credited in the captions.

Breeding catfish in tanks often leads to colour and form mutations. Pictured here is a Graceful Catfish, Ictalurus punctatus. Photo: Aqualife Taiwan

There is a wide variety of catfish species, some of which are now available in good aquatic centres. Among them are a large number of really interesting species, which are often extremely useful members of the aquarium, and which are generally called "cleaners". Aquarists take advantage of their habit of searching the floor of the aquarium for anything edible – for example, left-over food or small worms – to contribute to the cleaning of the tank.

If you are looking for colourful, lively fish, you will not find many among the catfish types. Certainly there are shoaling fish, like the Glass Catfish, but they are not colourful. Upside-down catfish or sucker-mouthed armoured catfish are more colourful, but they are often nocturnal or very shy. Nevertheless, many aquarists are fascinated by these secretive species and it is not without good reason that catfish are becoming increasingly popular in the hobby.

During feeding – catfish have a well-developed sense of smell – and at night time the aquarium becomes very lively. The experienced catfish aquarist will not miss the moment when the lights go off at night. With a torch or using very weak aquarium lighting, no more than a small 10W light bulb, every enthusiastic nature-lover will discover: catfish are simply fascinating.

Although most catfish are initially introduced to the aquarium as a

3

Many more catfish are bred for human consumption than for aquatic purposes. Occasionally some unusual types are exported for the aquatic trade – like this albino Clarias batrachus – but this is quite unsuitable for keeping in a home aquarium.

"working fish", soon many aquarists get quite enthusiastic about these entertaining fish. Sucker-mouthed armoured catfish, like members of the *Ancistrus* or *Otocinclus* genera, are used as algae eaters in the aquarium, whereas cory catfish are introduced to help clean the tank through their constant search for food. Careless purchases of unknown species can lead to problems. Be careful – there are quite a few types of thievish catfish which, even when young, are capable of decimating the stock of small fish in your tank.

As many catfish feed mostly at night, the reason for the apparently inexplicable disappearance of fish from the tank will not be immediately obvious. Never buy a catfish without knowing its eating and living habits. Also bear in mind that some catfish will grow to a considerable size, even though they look very pretty when young.

It is always difficult to interest people in particular types of fish as every aquarist has his or her own ideas and preferences. In their opinion only their favourites display unique colours, and are unrivalled in their appearance and fascinating in their behaviour. Naturally, all this also applies to catfish!

So, let yourself become fascinated by these diverting fish.

Many different combinations of larger catfish types and cichlids can be kept harmoniously in a tank. Photos: Dr J. Schmidt

If you are planning to keep several different species of catfish in your aquarium, make sure that you know all about their origins and their specific requirements in terms of water quality and temperature. As catfish come from a wide area and have different requirements, it is essential that their individual needs are recognized.

The aquarium

The size of the tank depends on the size of the adult catfish. This is important, as some catfish can to grow to several metres long. Whether you choose a framed tank or a simple silicone-sealed container is less important.

> **Tip: The best place for the aquarium is in a quiet spot, away from bright light, where the catfish are not startled by people carelessly walking past.**

Tanks which are too bright and with no hiding places can prove a stressful environment for catfish. A lounge or study is usually a better place than a corridor or kitchen. At the same time

The catfish aquarium

It is important that the aquarium provides at least one hiding place which should look as natural as possible. In this picture a pleco, Glypto-perichthys gibbiceps, is trying to hide from bright light. As sucker-mouthed armoured catfish pose no danger to other fish, it is possible to keep larger species together with other small fish. Photos: bede-Verlag

the catfish should get used to humans so that the slightest movement does not send them into hiding.

> **Tip: Tanks made of synthetic or acrylic materials are not suitable for keeping many types of sucker-mouthed armoured catfish as they scratch the surfaces with their rasping mouths.**

An aquarium for catfish must never be too small. Not only do many catfish grow quite large, some can live for 50 years or more. The minimum length of a tank should be 1 metre. Especially for sucker-mouthed armoured catfish, bogwood roots are essential in the tank. The plant stock depends on the number of catfish which are to be kept in the tank as many fish are more or less vegetarians. The majority of catfish

species are nocturnal. Sucker-mouthed armoured catfish tend to spend most of the day stuck by their sucker mouths to the same place in the aquarium, either under a piece of wood or rock. Many catfish are territorial over their resting place and will defend it against intruders.

Any wood put in the tank will be scraped by the catfish which releases

tannins and causes the water to turn yellow. While this affects water clarity, it is totally harmless.

> **Tip: Sharp edges on any rocks and stones should be smoothed with an angle grinder – wear protective goggles!**

In order to give catfish some privacy, the tank must have sufficient hiding places. Shelters can be made of bogwood or flat stones stuck together with silicone adhesive. But you can't use just any type of wood as this would rot in the warm water. Bogwood or Scottish moor oak is ideal and a large choice is available from aquatic shops. Larger pieces of wood must be fixed to the floor of the tank, for example with a heavy stone.

Apart from offering a hiding place, the wood also has another function; many sucker-mouthed armoured catfish constantly grate the wood with their mouths and at the same time ingest living organisms as well as cellulose, a substance which is essential for their survival.

> **Tip: In breeding tanks, clay or plastic pipes can be used as hiding places.**

The back wall of the tank must not be made of synthetic materials or cork if you want to keep sucker-mouthed armoured catfish as they would scratch and grate the material. For an attractive backdrop in an aquarium for large catfish, the aquarist has to be satisfied with a pretty printed background which can be glued to the back wall of the tank. Pieces of slate or other stones can be used as an alternative.

Substrates

Substrates must always be soft and have no sharp edges to prevent the catfish from injuring their delicate barbels while searching the bottom for food.

As many catfish prefer to live on the floor of the tank, the morphology of their underside makes it essential that only river sand or very fine gravel with a grain size of 2mm-4mm is used as a substrate medium.

Even large species of catfish, like this Phractocephalus hemioliopterus, need a soft substrate to enable them to dig without injuring themselves. Photo: bede-Verlag

Quartz gravel or hydrocultural clay bubbles are completely unsuitable, the first can cause cuts and lacerations and the latter is far too rough for the rummaging catfish.

> **Tip: Sand is not very good as substrate as its density prevents adequate aeration – small pockets of decaying matter can form which will be harmful to plants and subsequently detrimental to the fish's health. Fine-grained gravel or smooth granulate is more suitable for aquatic plants.**

In order to compromise to meet the requirements of both plants and the burrowing catfish – like many cory catfish or some dorads – the usual aquatic gravel can be used for the planted area and a flat bowl filled with sand placed in the front of the tank. Feeding will then have to take place above the bowl so that the catfish get used to it. The sand in the bowl can then easily be replaced if it begins to get clogged or dirty.

For the aquarist, catfish only ever start to lose their attraction when their normally active feeding habits are compromised by a dirty substrate. As many catfish have adapted to life on the floor of the tank and are almost constantly active, they deserve a clean substrate in which they can burrow freely. The height of the aquarium is not particularly significant and depends mainly on the size of the

Cory catfish, like this Aspidoras *sp., need a soft substrate so that they can burrow well without injuring their delicate barbels. Photo: Dr J. Schmidt*

catfish that it contains. The floor area, however, has to be large enough to offer all the catfish sufficient space in which to graze.

Studies carried out in natural biotopes have shown that catfish live almost exclusively on soft surfaces (sandy or muddy), which are covered with gravel, sunken wood or leaves.

> Tip: It is advisable to try to recreate a piece of a tropical riverbank in your aquarium.

For this you will need a mixture of fine and larger rounded gravel; gravel with sharp edges will injure the catfish's barbels. Open areas combined with densely planted zones make an interesting habitat for which *Anubias, Echinodorus, Cabomba, Heteranthera, Myriophyllum* and many other plants are suitable. Stone structures and bogwood complete the design and give the catfish a sense of security.

A group of cory catfish can be particularly dynamic. One minute they look as if they are playing, dancing around each other, settling on the plants only to disappear quickly into one of their hiding places and allowing order to be restored again. In other words cory catfish liven up your aquarium. Their continual activity also helps detritus to be wafted towards the suction valve of filter. Every now and then the hive of activity is interrupted by longer periods of stationary rest.

Filtration

Some catfish like to dig burrows underneath the fitments in the tank by whirling through the substrate which stirs up mulm. To avoid the water turning cloudy, a powerful filter must be fitted to remove it.

> Tip: A filter which can process two or three times the tank's volume of water per hour is usually sufficient for a normally stocked catfish aquarium.

The filter material should be easy to clean and it is advisable to use a pre-filter. Most of these pre-filters are made of foam-filled cartridges , which should be cleaned in lukewarm water on a regular basis to avoid destruction of bacteria that colonize them. Ideally the filtration system should not cause powerful water movement as some catfish do not like strong currents. On the other hand a certain amount of water movement is necessary for the well-being of the catfish.

> Advice: Despite filtration, some waste will remain in the water; so it is important to carry out a regular partial water change. Aim to change one third of the water every fortnight.

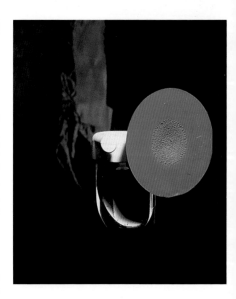

Use of constant tests like this enables you to control the pH value of the water, and indirectly the CO$_2$ level, over the long term. Photo: bede-Verlag

Harmful substances present in the tank are diluted by carrying out a partial water change. The amount of water to be changed depends on the tank size and the size of the fish population in the tank. For example a third of the water can be replaced every fortnight.

Water chemistry

Most catfish originating from the tropics prefer the same water quality as many tetra species, in other words soft and slightly acidic. Despite these requirements many aquarists keep them in hard (up to 25° dGH) and alkaline water. Although many catfish can adapt to these conditions, ideally the water should be soft to slightly hard, 4-20°dGH with a pH value of 7 or slightly lower.

> **Tip: The temperature can range between 20 and 27°C, but 24°C is ideal.**

Take care:
Fish from the Amazon lowlands usually require a higher water temperature! Catfish from these regions require a water tempera-ture of 25-30°C.

A powerful filtration system provides constant crystal-clear conditions. Nearly all catfish love flowing water, but not a strong current; this can be achieved with a power head if necessary.

General advice on what type of water is suitable for all catfish cannot be given because of their diverse origins. But the care of most catfish is easy and advice on looking after unfamiliar species can be found in the specialist literature (see Further Reading). Often normal tapwater can be suitable for catfish. However, for species imported from South America or Southeast Asia the water should never be harder than 15°dGH. As many succeeding generations have acclimatized to our water conditions, it is quite possible to keep them in soft tapwater. Consequently, it is not necessary to imitate the water conditions of South America. The pH value of very soft water is often not very stable and can be harmful to the fish.

> **Tip: Water with a hardness value lower than 4°dGH does not contain enough minerals to keep the pH value stable enough to satisfy fish and plants.**

In a newly established aquarium it is vitally important to check the water conditions regularly to ensure that the

the so-called nitrogen cycle has to be completed, which means that appropriate micro-organisms have to be established and allowed to multiply. Approximately two weeks after setting up, sufficient bacteria should be present to cleanse the tank of any fish faeces, food remains and dead plant material.

External filters have also proved to be useful in catfish aquariums. This type of filter can also be fitted with a heater. Photo: bede-Verlag

After two to three weeks the micro-organisms will have multiplied sufficiently to convert any harmful substances into less harmful ones. For example, any remaining nitrates in the water can be absorbed by plants and they are far less toxic to catfish than nitrites or ammonia.

pH value and the dissolved nitrogen compounds are not at levels which might be harmful to the catfish. The ideal pH value for keeping most catfish species is just below neutral. Occasional peat filtration, the addition of carbon dioxide or special substances which lower the pH value, all of which are available at aquatic shops, will help you to establish the correct pH value.

Tip: Regular partial water changes help to keep nitrates at acceptable levels – as long as the water does not already have a high nitrate content.

Tip: Every aquarist who alters the water conditions must check the pH value at least twice a week.

If there is no necessity to do so, then no further tests or experiments should be carried out with the water, although for breeding purposes it is sometimes necessary to change the water chemistry.

You should never introduce fish into a newly established aquarium. The tank has to be ecologically "run in" as

The water temperature in a catfish aquarium must not be too high. Around 22 to 26°C is a range in which catfish feel most comfortable, while some will tolerate even lower temperatures – down to 20°C. In nature water temperatures fluctuate, especially during the seasons, and many fish – not just catfish – do not survive these periods.

Such fluctuating temperature conditions must never exist in an aquarium, even though they would

mimic what occurs in nature. Many "nature lovers" who seem to think that fish should be kept under natural conditions do not realize that many of them do not always live in an ideal habitat in the first place.

Heating

Heating coils which are hidden in the bottom material are not advisable in catfish tanks, as the fish could dig them up and gnaw them. However, such heaters are suitable for tanks which house small catfish species like *Otocinclus* or *Aspidoras*. Tempera-ture-regulated rod heaters are generally fitted in catfish tanks; there are unbreakable versions which are ideal for the larger catfish species. Even better would be to have an external filtration system with a built-in heater.

Lighting

Most catfish are thought to be either nocturnal or only active during twilight. This assumption stems from the fact that many catfish are rarely seen in bright conditions.

Tip: In dim lighting conditions or if parts of the aquarium are shaded with floating plants, catfish are also active during the day.

It is sensible to introduce dim lighting for approximately one hour in the morning and evening before and after the main lighting is switched on to allow the catfish to feed in these preferred low-light conditions. Of course the food must be introduced to the tank either during or just before these periods.

Planting

As catfish tanks are not often brightly lit, only use plants which do not need a lot of light. In the wild catfish are rarely found in an environment which is densely populated with plants.

Tip: It is very rare to use plants for the aquarium that come from the catfish's natural habitat.

Far left: While young Glypto-perichthys gibbiceps keep the plants free from algae, as they grow, they will also view the plants as food.
Photo: Dr J. Schmidt

Top right: Cory catfish are not algae eaters. Photo: Aqualife

Left: In order to achieve optimum plant growth and to influence the pH value, all catfish enthusiasts should use a CO_2 fertilizer in their aquarium. Photo: bede-Verlag

As aquatic plants have a beneficial effect on the quality of the water by taking up the waste products of the metabolic process and stabilizing water conditions, it is important to include some plants in your catfish aquarium. Aquatic plants do not just fulfil a decorative purpose; they also have an important ecological function.

It does not matter if the aquatic plants do not come from the same place as the catfish. For example, *Cryptocoryne* species originating in Southeast Asia are ideal for an aquarium stocked with American catfish, as most cryptocorynes do not need much light. This also applies to the African *Anubias* species which also need very little light to thrive in the aquarium. This swordplant has relatively hard and leathery leaves and can put up enough resistance against the rasping teeth of sucker-mouthed catfish to survive and even flourish in the catfish aquarium. Java moss, *Vesicularia dubyana,* and the hardy

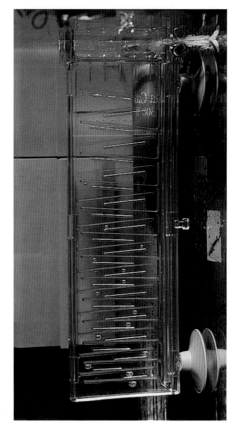

Beautiful aquatic plants, bogwood and catfish, here Sturisoma aureum, are often a successful combination. Photo: Dr J. Schmidt

Java fern, *Microsorum pteropus*, are generally well suited to a catfish aquarium. However, as the catfish constantly grate away at the substrate, these plants have to be fixed to the bottom so that catfish cannot dislodge them. Specialist shops sell a range of floating aquatic plants to shade parts of the aquarium. Even the little-loved common duckweed, *Lemna minor*, can be used for this purpose. A big disadvantage though is its growth rate and it will have to be removed from the water on a regular basis. It does, however, improve the quality of the water by absorbing large amounts of nitrates and other harmful substances produced by the fish.

Other suitable floating plants are the India fern, *Ceratopteris thalictroides*, and its relatives, as well as the butterfly fern, *Salvinia auricularia* and its relatives. They grow very well and are easier to remove than the duckweed as larger plants are much easier to fish out of the tank. Crystalwort, *Riccia fluitans,* is also a good floating aquatic plant although it is prone to be attacked by algae.

Finally, delicate plants like fanwort, *Cabomba, Limnophila*, parrot's feather, *Myriophyllum*, and even the robust Java moss, *Vesicularia,* are not suitable for a tank in which the catfish will root around in the substrate. Any stirred-up detritus will settle on the fine leaves and before too long the plant begins to look unattractive and it may even die.

General rule: With a few exceptions, aquatic plants with large smooth-edged leaves are ideal for a catfish aquarium.

Buying new catfish

From the outside most catfish do not display any symptoms of the many possible diseases that they may be suffering from. Therefore, buying catfish can be a risky business and it is important that the catfish enthusiast finds a reputable specialist shop in his or her area.

Tip: It is definitively worth paying a bit more to guarantee that your fish are fit and healthy.

If possible, try to avoid buying recently imported catfish. Effects of the stress caused by the transportation and any diseases only show after some time. Unfortunately, not many species of catfish have been bred in captivity yet and so they can only be bought as wild stock. In any case catfish should be held in a quarantine aquarium for at least two weeks.

Tip: Only when you are sure that the catfish are completely healthy should they be introduced into a tank with other fish.

Check the catfish for wounds and injured fins. The barbels and antennae must be whole and not show any sign of damage.

Catfish which constantly clamp their

fins are either ill or have suffered severe stress. Great caution should be taken before buying the fish. Also check their breathing, which should be calm and regular. Fast breathing is another sign of illness or stress.

Feeding

Although some catfish – especially the sucker-mouthed catfish – are described by aquarists as "algae-eaters", happily today they have lost their image as consumers of all the left-overs. If you keep many greedy and fast fish – those which don't even let food drop to the bottom of the tank – with catfish, you will find it difficult

Certain catfish, here Pseudo-doras niger, are often sold in shops as pretty little fish. As soon as they are in an aquarium, they will start eating all sorts of things, even other fish, and subsequently grow at a fast rate. This dorad was moved from a home tank to be rehoused in a zoo, but not every aquarist is lucky enough to find such a new home for their fish.
So – make sure you have all the relevant information before buying a fish!

Left: *When looking for food, catfish are very resourceful.*

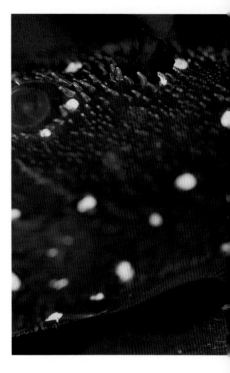

to care for the catfish, let alone to breed from them.

In some community tanks it may be possible to feed the catfish in a special area which other fish in the tank cannot intrude upon.

> **Tip: In most community tanks sufficient amounts of food fall to the bottom to feed the catfish.**

Unfortunately, most fish kept in tanks are overfed. Therefore, more of these fish die through being overweight than from underfeeding, though starvation is sometimes caused in exceptional cases by people with strict ideas on nutrition.

> **Tip: Moderate feeding will increase the life expectancy of your fish**

A good dry food can be the staple diet for most catfish. Food in tablet or granulate form is also suitable. However, algae-eaters or herbivorous fish must be fed at least twice a week – and when breeding even daily if possible – with fresh greenery, which can even be given frozen. Frozen live food is also suitable.

Most catfish like eating live food. Red and black mosquito larvae as well as water fleas are suitable. Krill or *Mysis*

are sometimes used, but they should only be fed sparingly because of their high salt content.

Although many aquarists dismiss them because of possible harmful effects – although unjustifiably so if used correctly – *Tubifex* worms are an acceptable food for many catfish species. These worms have to be soaked for several days before they can be fed to the fish.

Frozen food should not be allowed to defrost before feeding, but can be added to the tank while still frozen. No catfish has ever suffered an upset stomach caused by frozen food.

> **Tip: As soon as frozen food thaws, it begins to deteriorate.**

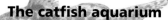

usually improved if the well-aerated hatching liquid does not contain too much salt, 8-10g/l is sufficient.

While older catfish can consume some eggshell – the remains are excreted undigested – it can cause obstructions in the intestines of young fish and may ultimately lead to death.

Tip: The hatched *Artemia* brine shrimps have to be carefully removed from the breeding container and thoroughly cleaned.

With every minute it is exposed to air unnecessarily, harmful substances already present increase steadily; they can be harmful to fish and affect the quality of the water adversely. In addition the vitamins in the food start to decompose quickly during the thawing process. The young of many catfish species can be fed immediately after hatching with *Artemia* brine shrimps. Only very few catfish have spawn that are so small that the hatching young have to be fed with infusoria and rotifers.

Artemia eggs available from specialist shops are of variable quality. They are also available in different sizes. It is important, that the eggs are fresh as the baby shrimps should have hatched after 24-48 hours. Hatching is

It is also important to rinse out as much of the salt water as possible. Then the *Artemia* brine shrimps are an ideal food for young fish and can initially form their main diet. Micro-worms and very fine soaked dried food – so that it can sink to the bottom – can also be fed to young catfish. If you have the opportunity to feed the finest pond food, even better. But you have to be ensure that no organisms are present in the food which might be harmful to the young catfish. For herbivorous and omnivorous catfish small peas are also suitable. While the catfish are very young, the pea skin should be split open so that the young fish can reach the soft inner core of the pea more easily. It goes without saying that the skins will have to be removed on daily basis from the

17

aquarium. In general most catfish should be fed very sparingly with animal-based food as the specialist herbivores will otherwise become overweight and could suffer from liver disease.

> **Good quality vegetable foods are:**
> • cucumbers or courgettes
> • carrots
> • blanched lettuce and spinach
> • blanched chickweed
> • blanched dandelion leaves
> • algae from flat stones
> • other soft plants

The aquarist has to test which food the catfish will eat. Even unusual food such as fresh peppers or cheese may be eaten by some catfish. Vegetable food remains can be left in the water for quite a while so that the food softens. Decomposing food remains have to be removed immediately. Many catfish are considered omnivores. They will eat peas, lettuce and dandelion leaves as well as mosquito larvae, *Tubifex* and all the tasty food pellets specifically formulated for catfish that are available from many specialist shops. As many catfish are nocturnal, it is sensible to feed the fish approximately half an hour before the lights go off in the aquarium. But one type of food is particularly popular with sucker-mouthed armoured catfish:

wood in any shape or form! The cellulose present in wood is actually essential for the general well-being of the sucker-mouthed armoured catfish, as their digestive system depends on it. Scottish moor oak or bogwood are hard, so the rootstock only dwindles very slowly under the constant rasping from the teeth of these catfish.

> **Tip: It is very easy to provide a varied diet for catfish in the aquarium. However, it is important to feed each species with the appropriate food.**

Cory catfish, for example, will eat almost anything edible. But one must not conclude from this that they exclusively feed on waste matter. They will eat any food left over by other fish in the tank but they also like all types of worms, like enchytraeid and Grindal worms and *Tubifex,* as well as small pieces of earthworms, which sink to the bottom of the tank. It doesn't matter if the food is alive or frozen. Cory catfish actively hunt for *Cyclops* and *Artemia* in the water and even catch fruit flies, *Drosophila,* on the water surface as well as taking pieces of flaked food. They do this swimming on their back, just like the Upside-down Catfish, *Synodontis nigriventris.* Some catfish even eat planarian flatworms and blue algae! You can enrich their diet every now and again with small amounts of cow's liver or shredded cow's heart; the latter is very popular with some catfish.

Diseases

If cared for properly, catfish are not particularly prone to diseases. However, occasionally some fish can become ill. To avoid this, newly acquired catfish must be kept in a quarantine tank for at least two weeks. This aquarium should be simply furnished and must have no contact with other tanks; all its equipment has to be kept separately.

Tip: The risk of diseases is especially high when new fish are introduced to an aquarium without undergoing the recommended quarantine period.

Bacterial diseases can usually be avoided. They are often the so-called weak parasites, which are only harmful to fish when the conditions in which they are kept are not ideal. A build-up of harmful substances, in particular, can cause bacterial disease.

Some of these bacterial diseases can be treated, and one of them is fin rot. A number of treatments are available from specialist shops, but they may only kill the bacteria but not eliminate the cause of the disease. Catfish suffering from dropsy, ulcers, bulging eyes or emaciation are difficult to treat and hardly ever survive. The affected fish should be removed from the tank and killed painlessly (for example, by freezing them). The cause of these symptoms is often fish tuberculosis. The bacteria causing this disease are actually found in almost every aquarium, but they do not harm healthy and happy fish.

Tip: As fish tuberculosis is not without risks to human beings, you should never put your hands into the aquarium if you have open cuts on them.

Occasionally, wild stock of catfish can carry the metacercariae parasite. If the fish are kept under the right conditions, the larvae will withdraw into a cyst and are then visible underneath the skin as light or dark spots. In this state they are no longer dangerous.

Sometimes catfish can have cloudy eyes which is often caused by unfavourable water conditions. The water must be improved immediately. In addition you can try treating the illness with human medicines by dabbing the eyes regularly with a cotton bud which has been soaked in a suitable preparation.

Also, anyone interested in keeping catfish should know that occasionally some aquarists have been stung by the fin spines of catfish and they have complained of a sharp pain which can last for several hours.

> **Tip: People suffering from heart disease are at real risk if stung by the fin spines of a catfish and must take all sensible precautions to avoid this.**

Extra care is called for when handling spiny catfish. Most of these stings happen when aquarists try to release catfish that have got stuck in a net by their fin spines. If threatened, catfish will spread their fins and literally lock their spines. The most effective way of removing the fish from the net – a technique which almost always works – is to place the net with the fish attached into a container filled with water and simply wait. Most catfish are able either to wriggle themselves free or escape from the folds of the net by retracting their fins.

The stomach of a healthy catfish is flat or slightly swollen, but never fat or hollow. The worst characteristic is a sunken stomach, which means that the fish has been starved or – even worse – is suffering from a parasite in the digestive tract. The bony formation of the dorsal area should be strong and always retains its shape. So, from above, you should get a clear idea of any swellings or physical abnormalities in your stock of fish.

Another important consideration when keeping catfish is never to treat them with any pharmaceutical preparations which could be poisonous or contain copper.

Sucker-mouthed armoured catfish rarely fall ill and are considered to be one of the most robust aquatic fish. This is certainly true of Liposarcus pardalis.
Photo: Dr J. Schmidt

Catfish, which belong to the sub-order Siluroidei and comprise many families containing over 2000 species, live primarily on the aquarium floor and have characteristic barbels on their heads and around their mouths. Catfish have adapted to a number of environments, they are not exclusively bottom-dwellers – some free-swimming species are found in the Siluridae family. Catfish use their barbels to assist them in finding food in murky water or after dark, which is when they are usually most active. With a few exceptions most catfish are freshwater fish. Only few species have made brackish water their natural habitat and even less are real marine fish like the Saltwater Catfish, *Plotosus lineatus*.

Breeding biology of catfish – Siluroidei

As with other large groups of fish there are several methods of reproduction known for catfish. From species that are egg scatterers which do not care for their brood, through plant and substrate spawners to varieties of cave- and mouthbrooders that care for their young, catfish show most types of breeding behaviour. Most do not care for their brood, or – if they do – usually the male does this.

The Glass Catfish, Kryptopterus minor, is a typical shoal fish. It is one of the few midwater swimmers and was previously known by the name Kryptopterus bicirrhis. Photo: Aqualife

Glass catfish, Siluridae and Schilbeidae

Most catfish in these families do not look after their young. Some care indirectly for their brood by hiding the spawn well and defending this territory. Others just spawn anywhere so the eggs float on the water surface and are eaten by other fish. Here the number of eggs laid will have to compensate for the vast losses that are inevitably suffered.

Among the species which do not care for their brood is the glass catfish family, the actual Glass Catfish *Kryptopterus minor* (previously known as *K. bicirrhis*) as well as their distant relatives the African glass catfish, such as *Eutropiellus buffei* (previously known as *E. debauwi*). From the few and incomplete reports available about these shoal fish and midwater swimmers, we assume that they are egg scatterers, and that their eggs develop either while in flotation or when attached to plants.

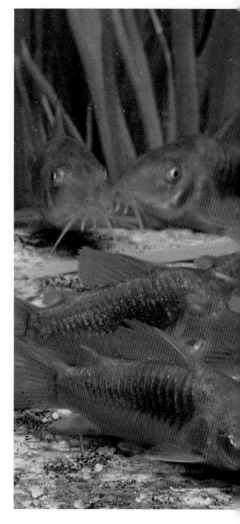

Wels Catfish, *Silurus glanis*

The Wels Catfish does, however, care for its brood. The male chooses a quiet area populated with plants where he fans out a hollow, which is then surrounded by plant material to create a nest-like structure. The male then takes it in turn with many females to spawn in this nest. After that, only the male will look after the eggs and later the hatched young, which look like tadpoles and initially stay in the nest. Once they get hungry and are eager to swim, the male is unable to keep them together and they "flee the nest".

Courtship and spawning of Corydoradinae

Even catfish which are very popular with aquarists, such as cory catfish like

Cory catfish include types that are easy-to-breed and some that are more complicated species. The Bronze Corydoras, Corydoras aeneus, is a particularly robust and popular aquarium fish. Photo: M.-P. & C. Piednoir

the *Aspidoras*, *Brochis* and *Corydoras* genera, do not care for their brood. In their natural habitat cory catfish mainly breed during the rainy season, which is during our Spring, i.e. April until June and sometimes from May to August, depending on the region. This characteristic is especially important when breeding from wild fish stock. The spawn only develops in slightly acid to neutral water conditions with a pH value of 6.2–7 and little hardness, between 3 and 10°dGH. For species from tank-bred stock, like *Corydoras aeneus*, *C. elegans* and *C. paleatus*, the time of year is irrelevant for breeding purposes. These species can breed theoretically all year round and the chemical characteristics of the water are less critical.

Catfish reproduction

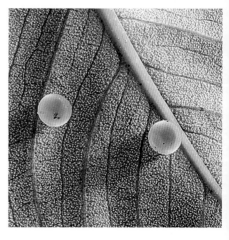

Cory catfish stick their eggs onto leaves or other items in the tank.
Photo: M.-P. & C. Piednoir

If the aquarist wants to breed *Corydoras*, then keeping a whole community is not just advisable, it is essential. You should keep two to three males for each female as a single male is normally not capable of fertilizing all the eggs. In addition you should not keep any other species in the aquarium.

As adult cory catfish do not care for their brood after mating, it is advisable to set up a breeding tank of 60-80 litres capacity (80 x 40 x 20cm). It should be placed in a quiet spot and ideally planted with large, stiff-leaved plants like *Anubias, Cryptocoryne* or *Sagittaria*. These plants should be secured between pebbles or under stone slabs with smooth edges. Substrate is not necessary at it would only gather unwanted detritus.

To stimulate the breeding partners you can carry out daily water changes of between 15 and 20 per cent. Spawning is then triggered by a 50 per cent water change, which reduces the temperature by 3-4°C. For the catfish this will seem like a heavy summer downpour, which causes their natural habitat to fill up. This is more than enough for easy-to-breed types like *Corydoras aeneus* and *C. paleatus* and leads almost automatically to spawning; some other varieties of catfish have also successfully been bred under these conditions.

Over a period of two to three days

Advice: You can find numerous valuable tips on breeding catfish in Catfish: Keeping and Breeding Them in Captivity *by Richard Geis.*

approaching the time of spawning the usually relatively quiet partners will become increasingly excited. You can watch the male rivals in pursuit of each other or confronting one another by ramming their heads together. However, this is nothing more than play and no-one gets hurt. After that the males start paying more attention to the females and try to hold on to one by touching her back with their barbels. If the female accepts this approach, they will mate for approximately 30 seconds during which – for most species – both male and female swim in the so-called T-position.

With every embrace the female expels between one and six eggs which she then gathers between her folded pelvic fins. There are a few differences among the species here: *Corydoras aeneus* produces approximately 20 eggs whereas the female of *C. hastatus* never produces more than one single egg. The eggs are then fertilized in the pouch formed by the fins after which the female then carefully sticks them

24

with her stomach onto a leaf or a wall of the aquarium.

Meanwhile the male will have released additional sperm into the water which increases the chances of successful fertilization even more. Both partners then take a short break.

In order to prevent the fish eating their eggs, it is essential to provide a constant supply of live food. Adding a disinfectant, available from specialist shops, to the water immediately after spawning will prevent fungus disease attacking the eggs. The eggs are distributed throughout the aquarium. Spawning can take from several hours to a couple of days, during which the males and the females regularly separate for some time. Meanwhile some of the first fry may have hatched which means that that the parents will have to be removed. Some species only produce few eggs (20-25 for *C. arcuatus* and *C. panda*), others can produce several hundred (up to 1200 for *C. zygatus*). The eggs are sticky and quite large. They require a temperature of 23-24°C for four to six days until they hatch.

Two or three days later the fry are swimming near the bottom of the tank. It is sensible to direct a spotlight onto the aquarium floor as this is where the live food – baby brine shrimps and micro-worms – are found. It is a well-known fact that the shrimps are attracted by light. Later the small

Corydoras will also eat *Cyclops*, Grindal worms and water fleas in suitable sizes. Keeping the aquarium meticulously clean is vitally important during this time and it is essential to remove any unwanted food and waste matter from the tank. Extra care should be taken when carrying out water changes as the young fry are very sensitive to changing conditions in their surroundings. The water conditions must not be altered significantly. It is therefore advisable only to change small amounts of water; the installation of a regulated drip-feeder would certainly be the ultimate. Once the young have reached a size of 15-20mm, they almost look like their parents. They will have developed a considerable robustness and can live for up to seven years.

Unfortunately, the really attractive cory catfish are often difficult to breed. Others, like these Corydoras aeneus, are easy to breed. Photo: M.-P. & C. Piednoir

Some peculiarities have been noted during the spawning of cory catfish. Although some leaves or stones are cleaned as possible substrate for the eggs during courtship – when the males chase the females and touch them with their barbels – the females almost always stick their eggs in completely different places. During spawning the catfish are much more unsettled. The whole shoal swirls around each other and the females are often approached and pressured by several males who touch them with their barbels.

Eventually one male will succeed in positioning himself crossways in front of the female and will grab her barbels with his pelvic fins. The female begins to lay her eggs into the pouch formed by her pelvic fins while the male simply releases his sperms in front of the female. The male then releases the female's barbels and the female swims through the sperm cloud with the eggs between her pelvic fins which are open at the front.

This is the reason why often only some of the eggs, especially the larger ones, of the cory catfish are fertilized. Finally the female decides on a suitable surface on which to stick her eggs. The eggs are not looked after by the parent fish, but nor are they eaten by the catfish as is so often the case with other species who do not look after their brood.

The Porthole Catfish, Dianema longibarbis, creates a bubble nest in which it lays its eggs. Photo: Dr J. Schmidt

Courtship and spawning of Callichthyinae

Undoubtedly one of the most interesting fish families is that of the Callichthyidae, which have distinctive anatomical characteristics: a strong, pointed head with several pairs of barbels on the jaw, generally large and movable eyes which is unusual in fish, a flat stomach and high-domed back, dorsal and pelvic fins with strong spines, and a skin which is not protected by scales but with bony plates which spread like a protective armour over its sides. These also curve over its head and back and so have given this fish the name of armoured catfish in the English language.

This is in aquatic terms the important family of the Callichthyidae, callichthyid armoured catfish which

are split into two sub-families according to their armour. They are the Callichthyinae, hoplo catfish which

include the genera *Callichthys, Dianema, Hoplosternum, Leptho-plosternum* and *Megalechis*. The second is the Corydoradinae family, the cory catfish which are split into three genera, namely *Aspidoras, Brochis* and *Corydoras*. The latter comprise most of the species in this sub-family. Although it is sometimes difficult to distinguish between them they represent the most interesting group aquatically.

The hoplo catfish of the genera *Callichthys, Dianema, Hoplosternum, Lepthoplosternum* and *Megalechis*, which are closely related to the cory catfish, take great care of their brood. The male builds a bubble nest on the water surface, preferably under a large plant leaf. Using a secretion from a gland in his mouth and atmospheric air, the male creates large bubbles which, mixed sometimes with plant particles, form the nest. Often the female also helps with the building of the nest. The pair will have moved into this territory several days earlier in order to get to know each other.

Spawning takes place in a similar way to that of the *Apidoras, Brochis* and *Corydoras* species, whereby the male grabs the female's barbels with the strong spines of his pectoral fins and clutches her tightly to his body. *Dianema* do not actually court before spawning. The female swims towards the male underneath the nest where she touches his genital area with her mouth and barbels while the male stays quietly in place. The male then hugs the female upon which she lays her eggs in the pouch made by her pelvic fins. The female sinks to the floor where she recovers briefly before swimming to the nest. Initially she touches the nest with her mouth, then rolls onto her back to stick the eggs to the bubbles, onto leaves or other parts of plants and onto other items in the vicinity – such as the walls of the aquarium.

In the course of several spawning sessions, the female can stick up to 700

Left: Dianema longibarbis.

Below: A picture of a male Leptho-plosternum pectorale. *All hoplo catfish build bubble nests although they are not necessary for the development of the eggs and brood.* Photo: Dr J. Schmidt

Ancistrus species are very popular amongst aquarists; they are sucker-mouthed armoured catfish. This is a female Ancistrus cf. hoplogenys. *Photo: Dr J. Schmidt*

eggs in place. Once she has finished spawning, the male chases her away. He will then care for the brood and also keep other fish away from the nest. The fry are watched for approximately four days until they hatch and sink to the bottom of the tank. The male still continues to watch the brood territory.

Life on the bottom for the young catfish

Typically the newly hatched fry of most catfish initially disappear into the cavities of the aquarium floor, where they will live for several days on the richly available micro-organisms found there, well protected from being washed around by currents in the water. The substrate also provides good protection from predators for the young catfish. Unfortunately, many

of the young catfish can get lost amidst large gravel on the aquarium floor as they cannot find their way out of this labyrinth. They either starve or suffocate. The grain of the substrate for a catfish aquarium should therefore be no larger than 2mm. This substrate is also ideal for many catfish as they particularly like digging around in the bottom of the tank. The young will reappear after about a week and start to form shoals or loose groups.

Sucker-mouthed armoured catfish, Loricariidae

Many sucker-mouthed armoured catfish of the Loricariidae family are popular with aquarists because of their brood care. Especially interesting is the sucker-mouth, with which they can attach themselves to surfaces.

These catfish usually live in fast-running or even torrential flowing waters where they rasp on stones and wood. The breeding habits of sucker-mouthed armoured catfish are only known with regard to a few of its species. Among sucker-mouthed armoured catfish there are examples of open breeders, substrate breeders and cave breeders. All of the known species care for their brood with the male being the carer.

Courtship and spawning of the Big-Finned Bristlenose

The best-known pattern of breeding behaviour is surely that of *Ancistrus dolichopterus,* a species which, given the right nutrition, will happily breed in any home aquarium. The male marks out a narrow hole and the surrounding area as his territory which he will defend against intruders. A female which is ready to spawn enters his territory and is initially chased away. Through her repeated attempts to

enter the territory, the male will eventually accept her and the pair begin touching each other with their mouths and barbels. Making obvious winding movements with his tail, the male leads her to the spawning place. Once in the cave the pair move behind each other and the female lays her eggs. After a while the male releases his sperm and with his fins fans them into the cave so that under normal circumstances all of the eggs are fertilized. After the female has attached all of her eggs in one or two batches she leaves the nest. The eggs are approximately 3mm in size and, typically for cave breeders, are bright orange. The male cleans the eggs with his sucker-mouth, after which he positions himself at the cave opening and begins to fan fresh oxygen-rich water in to the cave with the aid of his pelvic fins. Cave breeders' eggs which are not fanned in this way usually do not hatch. The eggs are generally

Ancistrus eggs are deposited by the female in one or two batches into a spawning cave made by the male.

Below left: *The male and female of the* Ancistrus dolichopterus *are only present together in the cave for the spawning. The male sits in the opening of the cave and fans his sperm towards the eggs. After spawning, the female leaves the cave and the male takes sole care of the brood.*
Photos: Dr J. Schmidt

A male
Ancistrus
dolichopterus
caring for his
brood.

Right: Regular
water changes
and plenty of
bogwood in the
aquarium
encourage
sucker-mouthed
armoured
catfish to breed.
Photos:
Dr J. Schmidt

fanned for four days until they hatch as fry with a large yellow yolk sac. The male blocks the cave's entrance by fanning it and so prevents the fry from escaping. After a further five days they will have eaten the yolk sac and the young catfish, only about 1.2cm in size, leave the cave for the first time looking for food. Only now does the motivation of the male to look after the brood subside. Usually neither parent follows their offspring. It has, however, sometimes been observed that the female will follow the yolk-rich eggs and the newly hatched fry which are flying out of the nest as a result of the fanning movements of the male. However, this only happens if the eggs, for reasons unknown until now, are not properly attached and the fry leave the cave too early.

Courtship and spawning of *Sturisoma*

The species of the interesting genus *Sturisoma* are open breeders with the male being the main carer. They are rarely seen in caves or hiding places other than during spawning. Their elongated bodyshape is a great camouflage between wood and plants. Members of the *Sturisoma aureum* species, who usually rest on bogwood during the day, become considerably more active at spawning time. The colour of the male turns darker and it chases the female through the tank. There is often more than one suitor and these males often compete against one another by lying next to each other, flapping their tail fins, or trying to push each other away head to head, as is also the case with other sucker-mouthed armoured catfish.

In the meantime the male keeps looking for a suitable spawning place where, once found, he lies, fins flapping as if already looking after the eggs. Soon the female approaches the fanning male and starts to clean the spawning place by rasping it with her mouth. The male also contributes to the cleaning of the substrate during which he attaches himself next to the female but often staying a little further back. According to Franke, the male's rostrum, the extension of his head, is level with one of the female's pelvic fins. For mating the male jerks forward so that in reverse the female's head is level with the pelvic fins of the male. The male now tries to impress the female by turning on his side until his back points to the female, his fins fully extended all the time. He then turns upright again whereupon the female sucks up to one of his pelvic fins. Shortly afterwards, with jerky pressing body movements, the female will lay three to seven eggs which stay attached a few centimetres behind the pair. The female lays her eggs with jerky movements while lifting her tail slightly. In the course of several spawnings – the whole process only lasts about half an hour – the female lays roughly 120 eggs which are attached to the spawning site somewhat randomly.

Sturisoma panamense has the same spawning pattern but its eggs stay close together and are attached in an orderly way in an oval pattern. The female of both species leaves the male after spawning while the male positions himself above the eggs and takes sole care of them. He lies above the spawn and fans it with his pectoral and pelvic fins constantly supplying fresh water to the eggs. The fanning movements become more intense towards evening. The eggs hatch after approximately five days in a water temperature of 26-28°C. The fry will have eaten their yolk supplies within a further three days and begin looking for food. The are now independent of their father and begin moving away from the shoal.

Substrate spawners and mouthbrooders

Catfish, like other larger sub-orders or fish families, are split into different brood-care types whereby the higher specialized cave- and mouthbrooders are morphologically and phylogenetically derived from the more primitive egg scatterers or substrate spawners. By looking at other species it becomes clear that there is a close connection between the different types of brood carers. A direct link can be drawn from the characteristics of the eggs, their quantities and their developing young. Egg scatterers lay numerous small, mostly camouflaged coloured or glassy eggs. As far as is known, glass catfish varieties fall into this category.

Substrate spawners that do not practise any brood care, e.g. the Corydoradinae, lay fewer larger eggs which, with their glassy to olive or white appearance, are camouflaged and are often attached in hidden places. Brood-caring substrate spawners, like the species of the subfamily Callichthyinae or the *Sturisoma* species, lay larger, but still camouflaged, olive or white eggs, and real cave breeders like *Ancistrus* and *Pterygoplichthys* species (in the broader sense including the genera *Glyptoperichthys* and *Liposarcus*) can afford to produce coloured, orange-yellow eggs, as they are well hidden away from any predators in the cave.

There are, however, exceptions. *Rineloricaria* species partly produce yolk-rich yellow or green eggs, which are, however, considerably smaller than those of the *Ancistrus* and the *Pterygoplichthys* species. One can assume that fewer eggs are lost if only a small number are laid as on average each pair has to produce two offspring per mating to safeguard the continued existence of the species.

Hiding breeders, to which mouth- and cave breeders belong, can lay fewer eggs which can be larger and more colourful. Larger eggs produce large fry, which are subsequently more robust and are able to catch larger prey and eat bigger food from an earlier stage. The bigger fry have no real competition and are ahead in the race to develop. The large eggs are of course more prominent and will have to be protected. The orange-red tinting of the yolks must be a real advantage for the development of the young, because everywhere else in nature where eggs are hidden or otherwise protected, the yolks are colourful. This does not only apply to cave-breeding catfish but cichlids, labyrinth fish and other families produce coloured eggs. The reader just has to think about the

Sucker-mouthed armoured catfish, here the Bronze Corydoras or Corydoras aeneus, belong to the open-spawning species that do not practise any brood care; the "cleaning" of the spawning substrate is nothing more than a ritual and is not done to keep eggs and substrate clean. Photo: Dr J. Schmidt

similar example of birds' eggs with their yellow yolks, which are often protected by a camouflage-coloured shell and experience intensive brood care. With development and progressive specialization of the species, the amount of spawn decreases and the egg size increases very gradually – with a few exceptions. For example, the small mouth-brooding cichlid *Pseudocrenilabrus* lays larger eggs than the enormous egg-scattering *Boulengerochromis*. However, the overall body size has to be seen into relation to the egg size, as in the nature of things *Silurus glanis*, which can grow up to 3m long, produces considerably larger eggs

than the small 4cm-long *Aspidoras* catfish. Most catfish eggs have a sticky secretion or a special clinging apparatus, which enables the catfish to stick their eggs to surfaces under water. In some cave- and mouth-brooders and breeding parasites (cuckoo catfish of the genus *Synodontis*), this clinging apparatus has strongly regressed. The special catfish characteristic whereby only the male cares for the brood avoids the need for complicated courtship games – as is the case, for example, with many species of perch – so that they can get to know their partners. Species that exhibit intensive brood care have developed sexual dimorphism in the

form of barbels or antennae which enable the partners to distinguish between the sexes, and at the same time they also probably serve to identify the species.

Mouthbrooders: shark catfish, Ariidae

Some genera of catfish are mouth-brooders, just as in other fish groups such as cichlids, Cichlidae, labyrinth fish, Anabantoidei, and snakeshead fish, Channidae.

Many members of the Ariidae family are mouthbrooders. Again it is the males who care for the eggs in their mouths. To leave the majority of the egg care to the males seems to be advantageous to these species as the females, weakened by the spawning, can eat to recover quickly and before too long start producing new spawn again. This makes them available for breeding much sooner and repro-duction time is shorter.

Here is a hypothesis: with cichlids, where females often care for the brood, the percentage of females in the overall population actually balances out numerically as the more numerous males are at greater risk from predators, because of their bright colouring. This explains why fewer females are born than males; it is mostly males who fall victim to predators. Subsequently several females spawn with one male and the brood is safer in the care of the female because of her camouflage. There are exceptions where the females are colourful, and in these cases often the males will take over the care or share it with their partner.

As many catfish are very prickly and armoured with bone plates, only very few of the well camouflaged and nocturnal species fall prey to other predators. Catfish are really only at danger while young, therefore the importance of intensive brood care cannot be overestimated.

Shark catfish are often marine fish, so very little is know about their breeding habits. According to Vogt the eggs of the Soldier Catfish, *Osteo-geneiosus militaris,* are yellow and approx. 1cm in diameter and the male only carries 10-15 eggs. Another species of the *Tachysurus* genus apparently carries eggs and fry in different stages of development in its mouth and even young fish up to 4cm long are still protected in the mouth. The species *Tachysurus sagor* can grow up to 50cm long.

Other species of the Ariidae family, which mainly live in the sea, are also predominantly mouthbrooders. Some of the *Arius* and *Hexanemichthys* species live in brackish water, others are only found in fresh water. Therefore, some of the species require the addition of sea salt to the water if kept in an aquarium. However, breeding these species has not yet been successful.

More is known about the shark catfish *Arius seemani*, which is very common in western America, and often sold in aquatic shops as a mini-shark. This species can grow up to 30cm long and is a mouthbrooder just like *Hexanemichthys bernayi* and *H. leptaspis*. These *Hexanemichthys* species originate in Australia where they are sometimes kept in tanks. Both species are very similar in appearance; *H. bernayi,* however, only grows to 25cm long whereas *H. leptaspis* can reach sizes of up to one metre.

Both species are mouthbrooders and the males carry approximately 150 1.5cm-large eggs in their mouths. During the spawning season the females develop a fold between their pelvic fins, which is probably designed to catch the eggs. Unfortunately, nothing else is known about their courtship and breeding habits.

From substrate- to mouthbrooders

With the limited information available today, very little is known about the development of brood care in the mouths of catfish. It can, however, be assumed that this way of looking after the young, in which eggs are held in their mouths in order to keep them clean, is the starting point of this development. With catfish it seems obvious that when fish do not practise brood care (or when both parents look after the brood), both sexes are more or less similar in colour and appearance. With progressive specialization in brood care where both sexes have different parental responsibilities– as, for instance, is the case with cave breeders – the fish display definite sexual dimorphism (as is the case with many livebearers).

Synodontis multipunctatus is a "cuckoo catfish".
Photo: M.-P. & C. Piednoir

Species

Big-Finned Bristlenose

Species: *Ancistrus dolichopterus* (VALENCIENNES, 1840) (13-15cm).

Origin: Amazon, North Brazil, French Guyana and Surinam.

Habitat: Like most sucker-mouthed armoured catfish, this species prefers the shorelines of still and flowing waters. Catfish are usually found where wood has fallen into the water; however, in their natural habitat Big-Finned Bristlenoses are not dependant on this and can certainly do without cellulose for a while.

Aquarium: For a pair or one male with several females, a tank of 1m length is sufficient. The aquarium must contain bogwood. If the catfish should damage the aquatic plant leaves while rasping the substrate, additional vegetable food should be given. Caves for hiding and possible breeding places are important.

Water: 22-26°C, pH value 5-7, 1-18°dGH.

Diet: Algae, food tablets, food left-overs from other fish in the tank (can be meat-based), and vegetables such as cucumber, salad, peas and dandelions etc.

Spawning: In hollows and caves, which have either been built in the tank or dug by the fish.

Breeding: As long as the water is not too hard and male and female are mature, breeding is almost unavoidable (see Breeding chapter, pages 29-30).

Features: A placid and sociable catfish which is suitable for all tanks with soft or medium-hard water.

Because of its changing colours and markings, different forms and age groups were described as separate species. For example *Ancistrus temminckii* is an often used synonym.

My Tip: I can recommend this popular species to every aquarist. The young catfish especially are active algae-eaters, which makes them ideal first inhabitants of a newly established aquarium to keep the continual growth of algae down. This makes them one of the most "useful" fish in aquatics and, moreover, a very interesting species.

The Big-Finned Bristlenose, Ancistrus dolichopterus, is a popular sucker-mouthed armoured catfish, which is sometimes sold as A. temminckii. Photo: M.-P. & C. Piednoir.

Living in a dimly lit cave has caused the eyes of this male Ancistrus cf. hoplogenys to go cloudy. Photo: Dr J. Schmidt

Aspidoras
pauciradiatus *is
a demanding
small catfish,
which has to be
kept in a shoal.
Only very
experienced
aquarists should
keep False
Corydoras.
Photo:
Dr J. Schmidt*

False Corydoras

Species: *Aspidoras pauciradiatus*
(WEITZMAN & NIJSSEN, 1970) (3cm).

Origin: Brazil, Rio Araguaia and the Rio Negro.

Habitat: Fine-sandy water in tropical forests, especially black water.

Aquarium: These sensitive small catfish should be kept in a species aquarium. They should only be kept with other small, slow-eating fish with light appetites. Fine sand for digging in and a well-planted area are important in the tank.

Water: 24-28°C, pH value 6-7, 2-12°dGH.

Diet: Food in tablet form, fine flakes and live food.

Spawning: These fish are substrate spawners that do not practise brood care.

Breeding: Difficult. Start a group ideally with a preponderance of male fish. After a few days of extra nutritious feeding, for example with mosquito larvae, several partial water changes carried out every day can instigate spawning. However, even then successful breeding is quite an exceptional occurrence and much rarer than for cory catfish. The eggs hatch after about four to five days and as soon as the young fry are able to swim free, approximately one day after hatching, they can be fed with freshly hatched *Artemia* nauplii and very fine flaky powder food.

Features: These gorgeous small fish are unfortunately very sensitive. The main anatomical feature distinguishing them from *Brochis* and *Corydoras* species is the skeletal structure and especially the skeleton of the skull.

The body markings of these Giraffe-Nosed Catfish are slightly different to the normal pattern. This is why they are described as Auchenoglanis cf. occidentalis. Photo: M.-P. & C. Piednoir

Giraffe-Nosed Catfish

Species: *Auchenoglanis occidentalis* (VALENCIENNES, 1840) (40-50cm).

Origin: Africa, Lake Victoria, Nile, Congo and other rivers and lakes.

Habitat: Prefers the edges of bodies of water with plenty of hiding places.

Aquarium: A large aquarium of approx. 150cm length is important, although this catfish is not very active. It also needs places to hide and a fine substrate for digging.

Water: 23-27°C, pH value 5-7, 4-18°dGH.

Diet: Food in tablet form, left-overs from other fish in the tank and especially animal food.

Breeding: Not known.

Features: Relatively peaceful for its size. Although this catfish is often sold for aquatic purposes, it is really too big for an aquarium.

Black Lancer Catfish

Species: *Bagroides macracanthus* (BLEEKER, 1854) (17cm).

Origin: Indonesia: Borneo and Sumatra; but not in Myanmar and Thailand.

Habitat: Mountainous freshwater rivers and streams.

Aquarium: These catfish are happy in tanks of 1m in length, although the water has to be very well filtered. Bogwood is not essential in the tank, and, if fresh, is quite unsuitable as the water for this species has to remain clear.

They are suitable for a community tank with peaceful, not too small companions. An ideal set-up would be in an Southeast Asian aquarium with loaches, barbs, rasboras and labyrinth fish.

Water: 22-26C, pH value 6-8, 4-24°dGH.

This beautiful Black Lancer Catfish is sometimes confused with Bagrichthys hypselopterus. Its correct scientific name is Bagroides macracanthus.

As a comparison: the real Bagrichthys hypselopterus. This catfish is also sometimes sold under an incorrect name. Illustrations: Dr J. Schmidt

Diet: Tablets, flakes and left-overs from other fish in the tank.

Spawning: Unknown; males, however, have a clearly visible white urogenital papilla in front of the anal fin.

Features: The catfish is often sold and featured in some books under the incorrect name *Bagrichthys hypselopterus*. This species, when fully grown, can reach a size of up to 40cm, twice the size of the real *B. hypselopterus* and as a predator it is hardly suitable for the home aquarium. Furthermore, *Bagrichthys hypselopterus*'s origin is Myanmar (Burma) and Thailand and it is only very rarely sold in specialist shops. Surprisingly enough, this catfish is also sometimes wrongly identified as *Heterobagrus bocourti* – a completely unnecessary confusion, which could easily have been avoided if original notes on this species had been studied carefully, especially as correct aquatic literature on the identification of this species is available.

My Tip: The real *Bagroides macracanthus* is an attractive aquarium fish and, if kept in the right conditions, would be worthwhile breeding.

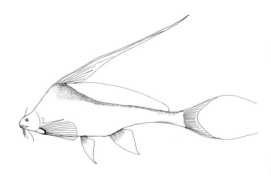

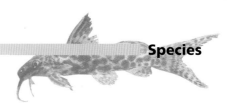

Brachyrhamdia marthae *lives in its native habitat in shoals with cory catfish. The young fish are similar to* Corydoras pygmaeus, *and their shoals intermingle in the aquarium. Unfortunately, no studies in their natural habitat are available.*
Photos: Dr J. Schmidt

Brachyrhamdia marthae

Species: *Brachyrhamdia marthae* SANDS & BLACK, 1985 (9cm).

Origin: South America, Peru.

Habitat: Not known as these catfish are only known from imports of uncertain origin.

Aquarium: A tank with a water volume of 100 litres is adequate for some catfish of this species. Although it is not a real shoaling fish, it should – just like the *Corydoras* – always be kept in a group. A community tank with cory catfish and medium-large tetras is a good option.

Water: 24-28°C, pH value 6-7, 4-15°dGH.

Diet: Food tablets; it also feeds on the leftovers from other fish food but this is not sufficient as a main diet. Animal food, such as mosquito larvae and water fleas, should form the main part of its diet.

Breeding: Not known.

My Tip: This is a good tank companion for cory catfish.

The Emerald
Catfish, Brochis
splendens, is
very similar to
the Corydoras
species and has
similar
requirements.

The Armoured
Catfish,
Callichthys
callichthys, and
Dianema
longibarbis
(pictured), are
robust fish and
attentive
parents. It is this
characteristic
which makes
them very
popular with
aquarists despite
their unpre-
possessing
appearance.

Emerald Catfish

Species: *Brochis splendens* CASTELNAU, 1855 (9cm)

Origin: South America: Brazil, Ecuador and Peru.

Habitat: These catfish are mainly found around the river bank regions of slow-flowing waters and flood areas.

Aquarium: Larger tanks from 160 litres upwards like the *"Corydoras*-type", i.e. with a fine sandy substrate for digging and a dense plant population.

Water: 24-28°C, pH value 6-7, 4-16°dGH.

Diet: Food tablets, flakes and left-overs from other fish food.

Spawning: Substrate spawners without brood care.

Breeding: Similar to that of the cory catfish; the sexes are easily identified by their comparative body size, the female can lay up to 300 eggs, a surplus of males is therefore advisable.

Features: This is a magnificent catfish, although it is much more demanding to keep in terms of water quality than other cory catfish.

Armoured Catfish

Species: *Callichthys callichthys* (LINNAEUS, 1758) (18cm).

Origin: The tropics of South America.

Habitat: The banks of slow-running and still waters with plenty of hiding places.

Aquarium: Tanks with a minimum length of 1 metre are ideal for one pair or a small group, as well as some community fish. Each of the catfish has to have at least one hiding place.

Water: 24-28°C, pH value 6-7, 4-18°dGH.

Diet: Food tablets, left-overs from other fish food and animal food such as mosquito larvae or water fleas. It will also eat puréed beef heart.

Spawning: Bubble-nest builder with the male caring for the brood (see pages 26-28 for further details).

Breeding: Simple.

Features: Very aggressive during the breeding season.

Headstander Catfish

Species: *Chandramara chandramara* (HAMILTON, 1822) (10cm).
Origin: India, Bangladesh and Myanmar (formerly Burma).
Habitat: Flowing and still waters, also found in poorly oxygenated swamps and paddy fields.
Aquarium: A tank of up to 80cm in length is adequate for this peaceful catfish.
Water: 22-26°C, pH value 5-7, 4-16°dGH.
Diet: Food tablets, flakes and a variety of frozen foods.
Spawning: Not known.
Features: Hardly anything is know about this recently imported catfish. A breeding attempt would be worthwhile as this species is peaceful and easy to keep.

Walking Catfish

Species: *Clarias batrachus* (LINNAEUS, 1758) (60cm).
Origin: Southeast Asia: India, Sri Lanka, Bangladesh, Myanmar (Burma), Thailand, Vietnam, Laos and Malaysia.
Habitat: Found in all water types even in very poorly oxygenated waters and paddy fields.
Aquarium: This catfish should only be kept in large tanks, but it is very undemanding.

The Headstander Catfish, Chandramara chandramara, *is well suited to aquarium living.*

Above: Head of an albino Clarias batrachus.
Photo: M.-P. & C. Piednoir

*Left: A young albino Walking Catfish.
Photos: Dr J. Schmidt*

Cory catfish of the genus Corydoras are sometimes difficult to identify. An example of mistaken identity is the pictured Corydoras trilineatus, which is often sold as C. julii. Photo: M.-P. & C. Piednoir

Water: 18-28°C, pH value 5-8, 4-25°dGH.

Diet: Omnivore and predator.

Spawning: The female lays the eggs in a hollow, the male looks after the brood. The female lays up to 1500 eggs. With a water temperature of 30°C the eggs will hatch after just one day.

Breeding: Because of its size, breeding in tanks has not yet been successful; however it is possible to breed this species in ponds without any problems.

Features: This greedy catfish can only be kept in company with other large fish. Just bear in mind that their mouths and stomachs can stretch and they will swallow much larger prey than you think their bodies can accommodate.

My Tip: Hands off these catfish!

Cory Catfish

Genus: *Corydoras* LACÉPÈDE, 1803.

As it is difficult to pick singular typical examples from this heterogeneous genus, and as whole books are dedicated anyway to cory catfish, the following is general information on these popular aquatic catfish.

The genus *Corydoras* is found in large areas of the tropics of South America and between Trinidad and Argentina. However, the majority of species are found in Amazonian waters. The habitat of catfish is very varied, but they are rarely found in very acidic black waters. They live in shallow, slow- and fast-flowing waters. In rivers and streams they prefer to live in coves where the water flow is calmer. During the rainy season they move to the flood areas where, again, they prefer the overgrown rims. More than 100 species belong to this genus, but only a third are regularly imported and just two dozen species are used for breeding. Practically almost every stretch of water in Latin America has its own *Corydoras* species or subspecies. Two to three populations, rarely more, sometimes live on the bottom together with a *Characidium* species, a tetra from the *C. fasciatum* group.

However, *Corydoras* are usually not forced to live in a community with foreign species, although it can happen. Often one *Corydoras* population varies from that in neighbouring waters.

To identify them – which is mainly done by colour and markings – is therefore anything but easy. This leads to some

unidentified fish being imported with known species; for this reason the precise location of the places where they were actually found can be relatively significant in identifying them. During the rainy season the small cory catfish find plenty of food and can concentrate on breeding. This all changes with the beginning of the critical dry season. The water level of rivers and streams drops by several metres; some stretches of water disappear altogether. Temperatures rise and the water heats up. Numerous fish will now suffer from lack of food and oxygen and subsequently die. However, our little catfish have an additional respiratory system which enables them to use oxygen from the air immediately above the water surface.

For this they shoot to the water surface, take a deep breath of air and return immediately the bottom of their waters. They swallow the air bubble, which moves into the digestive system and eventually into the lower intestine where numerous capillaries extend through the mucous membranes and ensure that these small fish can profit by absorbing oxygen from the air. The used air is then expelled through the anus and rises to the surface in form of small air bubbles.

Occasionally, hoplo catfish leave their wet habitat and with the aid of the stiff rays in their pectoral fins move across the ground. With high humidity they can cover considerable distances to neighbouring waters. This behaviour has not yet been noticed in cory catfish.

Almost all cory catfish (not necessarily the larger species) prefer life in colonies in flowing waters where they mainly live on the bottom. Some species, such as *Corydoras hastatus* and *C. pygmaeus,* are not restricted to living on the bottom and they often swim freely in the water.

Cory catfish race around like mice, gleaming and glistening in the sunshine. This is particularly obvious during twilight when they gather to search for food. They rummage through the fine substrate by ploughing through it with their mouths. Here they find a variety of hidden micro-organisms, such as micro-worms, tiny crustaceans, insect larvae etc. Sometimes they will dig with their pectoral fins into the substrate in order to get at a particularly choice delicacy.

Cory catfish are peaceful and easy-going fish. They just need to live in a group of at least five or six fish in order to feel comfortable. This group has to be introduced to the tank at the same time, but even then occasionally a confirmed loner will go his own way. The reason for this behaviour is not yet known. *Corydoras* are perfect companions for other small or large fish as long as they are left alone by them. As a rule the sexes can be identified reliably. The females are rounder and larger than the males, which keep their slim figure. Occasionally the dorsal fin of the male is a little bit longer and more pointed.

Porthole Catfish, Dianema longibarbis, are good and robust aquarium fish. Photo: M.-P. & C. Piednoir

Porthole Catfish

Species: *Dianema longibarbis* COPE, 1871 (12cm).
Origin: South America: Peru.
Habitat: Sandy clear- and black-water streams.
Aquarium: Larger tanks from 1m long upwards with a sandy substrate and dense plantation, similar to that for cory catfish.

Water: 24-28°C, pH value 6-7, 4-18°dGH.
Diet: Food tablets, also left-over food from other fish in the tank.
Spawning: Porthole Catfish are bubble-nest builders with the males caring for the brood.
Breeding: Relatively simply (see pages 26-28 for more details.)

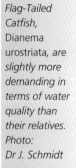

Flag-Tailed Catfish, Dianema urostriata, are slightly more demanding in terms of water quality than their relatives. Photo: Dr J. Schmidt

Flag-Tailed Catfish

Species: *Dianema urostriata* RIBEIRO, 1912 (15cm).

Origin: Brazil, Rio Negro.

Habitat: Sandy clear- and black-water lakes or tropical rivers.

Aquarium: Larger tanks from 1m upwards with sandy substrate and densely planted, similar to those for cory catfish.

Water: 24-28°C, pH value 5-7, 2-16°dGH.

Diet: Food tablets, flakes as well as frozen and live food.

Spawning: The males build bubble nests in which the females lay their eggs. Only the male cares for the brood.

Breeding: Feeding high-quality live food and regular water changes will encourage spawning. Breeding prospects are further improved by a slightly raised water temperature. The eggs and fry are cared for and defended by the male. The eggs hatch after three days and leave the nest after a further two days. This is when the male's parental care obligations diminish and he even considers the young catfish as possible prey. The fry can be fed immediately with freshly hatched brine shrimps and very fine flaky powder food.

Features: This species is a little more demanding than its robust relatives *Dianema longibarbis* and *Callichthys callichthys*.

My Tip: Initial breeding trials should be done with *Callichthys callichthys* or *Megalechis thoracata*.

Banjo/Frying Pan Catfish

Species: *Bunocephalus coracoideus* (COPE, 1878) (15cm).

Origin: The Amazon, Brazil, Argentina.

Habitat: Prefers to live on the bottom of calm waters.

Aquarium: A 100-litre aquarium is quite adequate for this inactive catfish. Hiding places are not necessary, although at least one part of the tank should have a dark, fine, sandy substrate.

Water: 24-28°C, pH value 6-8, 2-20°dGH.

Diet: Food tablets, left-overs from other fish food and frozen food. At least two weekly night feeds are necessary. This catfish accepts all types of food.

Spawning: The male provides the spawning hollow and cares for the brood.

Breeding: The female lays approx. 200 eggs. The tiny fry hatch after 24–36 hours and initially have to be fed with rotifers.

Features: These catfish are still sometimes known under the invalid genus name of *Dysichthys*.

A portrait of a Frying Pan Catfish, Bunocephalus coracoideus. These peculiar-looking catfish hardly move in the aquarium during the day and are therefore only recommended to specialist catfish enthusiasts. Photo: M.-P. & C. Piednoir

Spotted Pleco

Species: *Glyptoperichthys gibbiceps* (KNER, 1854) (50cm).

Origin: The Amazon, Peru.

Habitat: Prefers to live on wood in flowing waters, but is also found in still waters. Many catfish of this genus and their relatives follow piranha shoals and feed on animal food left-overs on the bottom.

Aquarium: Larger tanks from 2m upwards are essential for the correct care of this species. As these catfish require a high proportion of cellulose in their diet, bogwood is a must in the tank. During the long-term care of this species, the aquarist can watch the bogwood slowly disappear.

Water: 24-29°C, pH value 5-7, 2-18°dGH.

Diet: Feeds on plants as well as animal food and will even eat dead fish.

The Swallow-Tail Glass Catfish, Eutropiellus buffei, *is regularly imported. However, breeding is still quite difficult. Photo: Dr J. Schmidt*

The attractive Spotted Pleco, Glyptoperichthys gibbiceps, *is unfortunately quite a large fish. Photo: bede-Verlag.*

Swallow-Tail Glass Catfish

Species: *Eutropiellus buffei* GRAS, 1961 (8cm).

Origin: The Niger area of Nigeria.

Habitat: Clear and black flowing water.

Aquarium: A tank of 100 litres is adequate for a small shoal of six to ten catfish.

Water: 22-26°C, pH value 5-7, 4-15°dGH.

Diet: Flakes and granulates as well as live and frozen food.

Spawning: Egg-scattering species that does not practise any brood care.

Breeding: Successful breeding is exceptional and ideal conditions (rainy season) are not yet known (see also pages 21-22).

Features: This catfish was often mistaken for the rarely imported *Eutropiellus debauwi* species; *Eutropiellus vandeweyeri* is a synonym.

My Tip: A pretty catfish, ideal as a companion for tetras.

Spawning: Cave breeders like their close relatives *Liposarcus multiradiatus*.

Breeding: Breeding in an aquarium has not yet been successful. In their natural habitat and in large breeding lakes in Florida, the male digs hollows in steep clay banks where the eggs and the fry are looked after for a few days.

Features: Up until a few years ago catfish of *Glyptoperichthys, Liposarcus* and *Pterygoplichthys* genera were grouped together in the *Pterygoplichthys* genus. They were also mistaken for the *Hypostomus* genus and the *Plecostomus* synonym. However, their dorsal fin is considerably shorter (approx. 7-8 fin rays, while *Glyptoperichthys* and its relatives of both genus have 10-13 fin rays).

My Tip: The young of this genus and its relatives are regularly sold in specialist shops as they are keen algae-eaters. However, they grow quickly and once they reach a size of around 20cm they will also eat the aquarium plants. So unless you can transfer the growing catfish to a larger, more suitable tank, you would be well advised not to acquire them at all.

Dwarf Anchor Catfish

Species: *Hara jerdoni* (DAY, 1870) (4cm).
Origin: India, Nepal, Bangladesh.
Habitat: All water types.
Aquarium: Tanks from 60cm upwards are adequate. This nocturnal catfish prefers dense planting. An aquarium installation set up as for cory catfish is ideal.
Water: 12-26°C, pH value 5-7, 4-18°dGH.
Diet: Food tablets, flakes, left-overs from other fish food as well as frozen and live food.
Spawning: Not known.
Features: This small catfish of the Sisoridae family is quite remarkable when you consider that some of its relatives grow up to two metres in length.
My Tip: A recommended species for a species aquarium as well as a Southeast Asia community tank.

Hara hara is a robust, cute little catfish; although nocturnal, if kept with the right tank companions, it will search for food during daily feeding times. It is a peaceful catfish and makes a good companion for rasboras or labyrinth fish. Photo: M.-P. & C. Piednoir

The Dwarf Anchor Catfish, Hara jerdoni, only grows to 4cm, while its relative Hara hara reaches a length of up to 7cm. This species is also well suited to the home aquarium.

A young Ornate Pim, Pimelodus ornatus. Keeping this species with small fish is problematic as the catfish can regard them as food (see also page 66). Photos: H. Hieronimus.

The Spotted Woodcat, Auchenipterich-thys longimanus, belongs to the family of driftwood catfish. It grows to c.15cm long and considers very small fish as food. Otherwise this catfish requires the same care and conditions as fish of the Hassar genus.

The same advice applies to the Midnight Catfish, Auchenipterich-thys thoracatus, which only grows to 11cm in length.
Photos:
H. Hiernonimus.

Hassar iheringi *is only comfortable in the company of peaceful fish. Unfortunately, this catfish is very delicate and susceptible to diseases. Photo: Dr J. Schmidt.*

Hassar iheringi

Species: *Hassar iheringi* (FOWLER, 1941) (8-10cm).

Origin: South America, Brazil, Peru.

Habitat: Prefers tropical clear and flowing waters, but is rarely seen in flooded areas.

Aquarium: A medium-sized aquarium of 160 litres is ideal. These catfish need hiding places but they certainly should not be made out of bogwood.

Water: 24-28°C, pH value 5-7, 4-18°dGH.

Diet: Food tablets, left-overs from other fish food as well as a variety of live and frozen food.

Spawning: Not known.

Features: A peaceful catfish which can become very shy if kept with the wrong companions.

My Tip: Best kept with small tetras and, for example, a group of cory catfish.

Asian Stinging Catfish

Species: *Heteropneustes fossilis* (BLOCH, 1797) (35cm).

Origin: Large areas of Southeast Asia: India, Sri Lanka to Thailand.

Habitat: Inhabits all water types. Asian Stinging Catfish are also found in very warm waters and those lacking oxygen, such as shallow ditches and paddy fields.

Aquarium: Large tanks from 1.5m long upwards; this catfish is mostly nocturnal.

Water: 22-26°C, pH value 5-7, 4-18°dGH.

Diet: Food tablets, left-overs of other fish food, but also small fish, worms etc.

Spawning: Breeding has been repeatedly successful. Breeding from almost fully grown catfish is quite simple.

Features: Handle with care – the spines are venomous and can deliver a dangerous sting. The closely related black *Hetero-*

The Asian Stinging Catfish, Heteropneustes fossilis, is inconspicuous, rapacious, greedy and nocturnal – hardly an "ideal" aquarium fish.

pneustes microps species only grows to 20cm, and is more peaceful and therefore better suited for aquatic purposes.

The Small Asian Stinging Catfish, Heteropneustes microps, is more suitable.
Photos:
Dr J. Schmidt

The body colour of the Dwarf Hoplo, Leptho-plosternum pectorale, becomes a bit monotonous as it matures. This catfish is very robust and is also suitable for a community tank.
Photo:
Dr J. Schmidt

Port Hoplo

Species: *Megalechis thoracata* (CUVIER & VALENCIENNES, 1840) (10cm).
Origin: Large areas of the South American tropics.
Habitat: Muddy still and slow-flowing waters, but especially swamps. Prefers swamps and waters with heavy vegetation but can also cope with a certain amount of water pollution in the vicinity of human settlements, which it particularly seeks out when searching for food.
Aquarium: For a small shoal with a minimum of six catfish, a tank of 1m upwards is essential. The catfish need areas for digging but also places to hide and retreat into.
Water: 24-28°C, pH value 6-7, 4-18°dGH.
Diet: Food tablets, granulates, flakes and left-overs from other fish food. In addition it needs a balanced variety of live and frozen foods.
Spawning: The male builds a bubble nest and cares for the brood. Reproduction is similar to that of *Callichthys* and *Dianema* (see pages 26-28).
Breeding: It is ideal to keep a breeding pair permanently in a species aquarium. Readiness to breed can be stimulated by a good balanced diet and regular partial water changes. The male builds a bubble nest with bits of vegetation at the water surface. The female sticks the eggs to the vegetation. The female should be removed from the aquarium immediately after spawning so that the male can concentrate on the brood care. The eggs hatch after three days and rearing is easy.

Ghost Catfish

Species: *Kryptopterus minor* ROBERTS, 1989 (12cm)

Origin: Southeast Asia: Borneo

Habitat: Flowing waters of the tropics.

Aquarium: As this catfish has to be kept in a shoal, the aquarium has to be at least 1m long. Dense vegetation is important as the catfish are easily startled and could eventually die of shock.

Water: 24-28°C, pH value 5-7, 4-18°dGH.

Diet: Food tablets, flakes, frozen and live food.

Spawning: Egg scatterers that do not practise brood care.

Breeding: Rare, but possible.

Features: For many years this catfish was mistaken by aquarists for its close relative *Kryptopterus bicirrhis*. This catfish is, however, non-transparent and grows to almost 20cm – twice the size.

False Bumblebee Catfish

Species: *Pseudomystus stenomus* (VALENCIENNES, 1839) (15cm).

Origin: Indonesia and Malaysia: Borneo and Sumatra.

Habitat: Around the banks of slow-flowing waters.

Aquarium: Tanks from 1m upwards are ideal. This nocturnal catfish needs several hiding places so that it can choose its favourite location.

Water: 20-26°C, pH value 6-8, 4-25°dGH.

Diet: Food tablets, flakes, left-overs from other fish food and animal feed such as mosquito larvae.

Spawning: Not known.

Features: The spiky fins of this catfish get caught easily in a net. Although this happens with other catfish, it is especially likely with this species and it can get injured if it is not removed very carefully.

Ghost Catfish, Kryptopterus minor, are typical shoal fish.
Photo: M.-P. & C. Piednoir

The False Bumblebee Catfish, Pseudomystus stenomus, is not particularly attractive but is easy to keep in an aquarium.
Photo: H. Hieronimus

South American Bumblebee Catfish

Species: *Microglanis iheringi* GOMES, 1946 (6cm).

Origin: Venezuela and Colombia.

Habitat: All waters.

Aquarium: Although this species can be kept in tanks up to 60cm in length, it is preferable to keep a small group in a large tank. Otherwise easy to keep.

Water: 24-28°C, pH value 6-7, 4-18°dGH.

Diet: Food tablets, flakes, left-overs from other fish food as well as frozen and live food.

Spawning: Not known.

Features: There are several similar species of this genus, all of which stay relatively small, so a possible misidentification would not matter much as all of them are suitable for aquatic purposes.

The Electric Catfish, Malapterurus electricus, *is not suitable for ordinary tanks. Photo: H. Hieronimus*

The South American Bumblebee Catfish, Microglanis iheringi, *should be kept in a shoal, although it is not strictly a shoaling fish. Photo: Aqualife Taiwan.*

Electric Catfish

Species: *Malapterurus electricus* (GMELIN, 1789) (100cm).

Origin: Large areas of Africa.

Habitat: All waters.

Aquarium: Large tanks – for young Electric Catfish – from 2m upwards.

Water: 23-28°C, pH value 5-7, 4-25°dGH.

Diet: Omnivores.

Spawning: The females dig hollows in the muddy substrate where they spawn with the males.

Breeding: Not known.

Features: The electric organ of the Electric Catfish, a nerve, lies on the sides of the body between the skin and the muscles. The electric discharge can be harmful to humans.

My Tip: Not suitable for the home aquarium.

The Asian Red-tailed Bagrid, Hemibagrus nemurus (left and below), is often mistaken for other similar-looking catfish. It can grow to around 60cm in length in large aquariums.
Photos: Dr J. Schmidt

Asian Red-Tailed Bagrid

Species: *Hemibagrus nemurus* (VALENCIENNES, 1839) (60cm)

Origin: Southeast Asia: Malaysia, Thailand, Singapore and Indonesia.

Habitat: All waters, even brackish water.

Aquarium: A tank with a capacity of 1000 litres or more is sensible, as this catfish can grow quite large.

Water: 23-27°C, pH value 6-7, 4-18°dGH.

Diet: Food tablets, flakes, left-overs from other fish food, especially animal feed.

Spawning: Not known.

Features: This catfish is often mistaken for other species, some of which grow even larger than this type.

My Tip: Only consider buying this catfish if you are absolutely confident about the identification. Other species often outgrow the aquarium.

Although Oticinclus hoppei *is not particularly colourful, it is very popular with aquarists because of its interesting behaviour and especially because of its appetite for algae.*
The small suckermouth catfish of the Otocinclus *genus are often introduced to planted-up community tanks as algae-eaters.*
Unfortunately, they are delicate fish, which can only be kept with peaceful companions.
Photos:
Dr J. Schmidt

Otocinclus hoppei

Species: *Otocinclus hoppei* MIRANDA-RIBEIRO, 1939 (5cm).

Origin: Brazil, Bolivia and Peru.

Habitat: In clear waters with plenty of vegetation.

Aquarium: Small tanks are adequate, although immaculate water conditions are essential.

Water: 24-28°C, pH value 6-7, 2-12°dGH.

Diet: Algae and other plant food, but will also eat small amounts of brine shrimps or similar food.

Spawning: The eggs are deposited freely onto plants or other objects, similar to *Corydoras.*

Breeding: Very difficult, but definitively possible.

Features: These catfish are particularly peaceable and prefer to live in large groups. Although they are not real shoaling fish, they love the company of their own species. This must be borne in mind for aquatic purposes and so that they feel really comfortable, they should never be kept in smaller groups than six fish.

My Tip: If you keep soft-water fish like cichlids and rasboras, then these catfish are perfect for dealing with algae. This catfish is really only suitable for the experienced fishkeeper.

found especially in fast-flowing, clear-water rivers and streams. The rocks are covered with a thicket of algae. Its close relative, *Otocinclus macrospilus*, has been known to science since EIGENMANN & ALLEN described it in 1942. It was first imported live into Europe as early as 1911.

Otocinclus hoppei's main feature is its medium-long slim body. By comparison its head is quite large and flat with the mouth underneath with seam-like lips. Upper and lower jaws are equipped with small teeth which make a good rasping instrument. They can be moved like levers of a typewriter whereby the algae are pushed against the substrate and then torn off. The lips are equipped with small papillae and form a disc which the little fish uses to attach itself to any solid matter and thus resist the flow of the currents. The completely flat stomach and the low tail shape are particularly useful for this purpose because of their streamlined form, which is not dissimilar to that of tadpoles. Scales are replaced by bony plates, just like roof tiles. The lack of scales is a characteristic of all catfish. These plates also cover the head and make a good armour. The stomach, however, is not armoured and therefore is a weak point, the Achilles heel of *Otocinclus hoppei* as it were.

Otocinclus hoppei has the same care requirements as its relatives. It needs clear and clean water and algae, algae, algae....
Photo: M.-P. & C. Piednoir

The small suckermouth catfish of the genus *Otocinclus* and its close relatives cover more than 30 species, of which only about half a dozen are sold in specialist shops. The all have one common characteristic, namely they are effective algae-eaters. This genus belongs to the large family of Loricariidae.

Otocinclus hoppei comes from Southeast Brazil, more precisely from the wider areas of Rio de Janeiro, where it is

Asian Shark Catfish

Species: *Pangasius hypophthalmus* (SAUVAGE, 1878) (100cm).
Origin: Thailand.
Habitat: All waters.
Aquarium: Large tanks from 2m upwards.
Water: 22-26°C, pH value 5-7, 4-25°dGH.
Diet: Omnivore.
Spawning: Not known.
Breeding: In its home large numbers are bred in lakes.
My Tip: The cute young Asian Shark Catfish is still often sold in specialist shops. This species is only suitable for the experienced aquarist.

Pareutropius longifilis

Species: *Pareutropius longifilis* (STEINDACHNER, 1916) (8cm).
Origin: East Africa, Tanzania.
Habitat: Still waters.
Aquarium: The aquarium has to be quite large – at least 1 metre long – as this species has to be kept in shoals. Dense planting in the tank is important for the catfish to feel secure and comfortable in the aquarium habitat.
Water: 22-27°C, pH value 5-7, 4-18°dGH.
Diet: Food tablets, left-overs of other fish food as well as frozen and live food.
Spawning: Not known.

*The Red-tailed Catfish, Phractocephalus hemioliopterus, grows up to 1m long and is a predator, which does not make it suitable for home aquarium purposes.
Photo above: Dr J. Schmidt.
Photo left: M.-P. & C. Piednoir*

Red-Tailed Catfish

Species: *Phractocephalus hemioliopterus* (BLOCH & SCHNEIDER, 1801) (120cm).
Origin: Large areas of the tropics of South America.
Habitat: In hollows of larger flowing waters; all water types.
Aquarium: Very large tanks from 3m upwards.
Water: 24-28°C, pH value 6-7, 4-18°dGH.
Diet: Carnivorous.
Spawning: Not known.
Features: Although not very active, these very large catfish are hardly suitable for home fishkeeping as they need to be kept in very large tanks. Red-tailed Catfish are highly predatory, feeding on crabs, crayfish, fishes and even stingrays as socialization trials in show tanks have proved.

Physopyxis lyra

Species: *Physopyxis lyra* COPE, 1871 (4cm).
Origin: In the Rio Ambyiacu in Brazil.
Habitat: River banks with plenty of hiding places.
Aquarium: Small well-planted tanks from 60cm upwards with plenty of hiding places are suitable.
Water: 24-28°C, pH value 5-7, 4-18°dGH.
Diet: Food tablets, flakes, fresh frozen and live food. This catfish cannot survive on leftovers from other fish food alone.
Spawning: Not known.
Features: These nocturnal and relatively inactive catfish have to be fed at night so that they don't starve.
My Tip: Only suitable for catfish experts. These catfish should only be kept in species tanks to safeguard their diet.

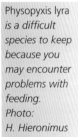

Physopyxis lyra *is a difficult species to keep because you may encounter problems with feeding.*
Photo:
H. Hieronimus

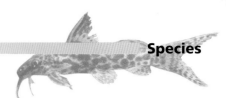

An unidentified small pim from the Pimelodella genus. Sometimes the newer genus name Brachyrhamdia *is* used, but no adequate differences are known so that for the time being the names from their first descriptions are used. However, this will change in the future.
Photo:
H. Hieronimus

Graceful Pim

Species: *Pimelodella gracilis* (VALENCIENNES, 1840) (15-20cm).

Origin: The Amazon in Brazil and Venezuela.

Habitat: In all areas of warmer clear- and black-water streams.

Aquarium: These lively and fast-growing pims need large tanks from 300 litres upwards. Although predominantly nocturnal, when kept in an aquarium they soon realize that they are in no danger and will eventually come out of their hiding places during the day – especially at feeding time. It is essential that each catfish has its own hiding place even though they sometimes share them.

Water: 23-27°C, pH value 6-8, 4-20°dGH.

Diet: Food tablets, flakes and left-overs from other fish foods.

Spawning: Not known.

My Tip: An aquarium with densely planted edges is advantageous for the fishkeeper, as it makes these Graceful Pim catfish feel more secure.

Above: *A* Rhamdia laticauda *from Honduras.* **63**

White-Spotted or White-Striped Pims, Pimelodus albofasciatus, *are – depending on their origin – different in colour. They can have a distinct stripy pattern or the stripes dissolve into a spotted line.*
Photos:
Dr J. Schmidt

White-Spotted or White-Striped Pim

Species: *Pimelodus albofasciatus* MEES, 1974 (25cm).
Origin: South America, in Guyana and Surinam.

Habitat: Lives in all areas of tropical flowing waters but prefers clear water.
Aquarium: These catfish – particularly active at night – need to be kept in large tanks of 300 litres or more. With extensive water changes and a good diet, these catfish become quite active during the day.
Water: 22-26°C, pH value 5-8, 4-20°dGH.
Diet: Food tablets, left-overs from other fish food, animal food such as mosquito larvae, worms and water fleas, as well as small shrimps. They even consider small fish as food.
Spawning: Not known.
Features: These pims have a variety of markings and therefore are available in many different colours in specialist shops.
My Tip: No matter how attractive a shoal of these young catfish looks in the shop, think very carefully before buying them as they grow very quickly, are greedy feeders and can live longer than ten years.

Dusky Pim

Species: *Pimelodus blochii* VALENCIENNES, 1840 (20-30cm).

Origin: Large areas of Middle and South America from Panama to Brazil.

Habitat: Lives in large rivers and streams as well as their tributaries. The Dusky Pim prefers clear and black water but they are also found in white water. During the day they hide on the banks and at night go hunting for food in large shoals.

Aquarium: These active pims need to be kept in large tanks measuring from 150cm in length with plenty of hiding places, where they can hide away from any bright light especially when first introduced.

Water: 20-26°C, pH value 5-8, 4-20°dGH.

Diet: Food tablets, granulates, left-overs from other fish foods and animal food such as mosquito larvae, worms and water fleas, as well as small shrimps. It will also consider small frogs as food.

Spawning: Not known.

My Tip: Although the different species of pims are sometimes sold under the name "Angel Cats", they are certainly no angels in terms of their behaviour and they must not be kept with small fish like, for example, neon tetras!

Photo below: Aqualife Taiwan

Above: *A Mystus cf.* vittatus *from Asia.*

Frequently some pims find their way into the shops which cannot be assigned to a species. Pictured here is Pimelodus pictus in its mature markings.
Photo: bede-Verlag

The Ornate Pim, Pimelodus ornatus, grows even larger than the other pim species imported for aquatic purposes and so is only suitable for a catfish expert who has sufficiently large tanks in which to keep it.
Photo: Aqualife Taiwan.

The Angelicus
Pim, Pimelodus
pictus, grows to
15cm and is
therefore well-
suited to a home
aquarium.
Photo:
Dr J. Schmidt

This pim also
eats small fish.
Photo: Aqualife
Taiwan.

The Chocolate Doradid, Platydoras costatus, is one of the few doradids that, because of its size, is suitable for aquatic purposes. It also poses no danger to other fish. Photo: M.-P. & C. Piednoir

Chocolate Doradid

Species: *Platydoras costatus* (LINNAEUS, 1766) (20-25cm).

Origin: The Amazon and Peru.

Habitat: Banks and areas full of wood debris in clear- and black-water streams, as well as quiet tributaries.

Aquarium: The Chocolate Doradid needs a sandy substrate or a sandy area in the aquarium as it likes to dig occasionally. If the roots of the aquatic plants are protected by stones, the doradid will leave them alone. In addition the fish needs a hiding place, for example under bogwood or a cave made out of stacked-up stones.

Water: 24-28°C, pH value 6-7, 2-20°dGH.

Diet: Food tablets, flakes, granulates, leftovers from other fish foods and only small amounts of vegetable food. Of course, animal food, such as live mosquito larvae, *Tubifex* worms or water fleas, is essential.

Spawning: Not known.

Features: The Chocolate Doradid can make growling noises which can sometimes be heard outside the aquarium, especially when it is disturbed.

My Tip: The Chocolate Doradid belongs to an entertaining species. They are very popular with catfish enthusiasts because of their unusual shape.

The Tiger Shovelnosed Catfish, Pseudoplatystoma fasciatum, can grow to more than 1m long and so, like all of its relatives, it is not suitable for aquariums in the home.
Photo: Dr J. Schmidt

An Upside-Down Catfish, Synodontis nigriventris.
Photo: M.-P. & C. Piednoir

Tiger Shovelnosed Catfish

Species: *Pseudoplatystoma fasciatum* (LINNAEUS, 1766) (100cm).

Origin: Paraguay, Peru, Venezuela.

Habitat: In almost all larger streams of the tropical forests in northern South America.

Aquarium: From a young age this catfish needs an aquarium of at least 2m in length.

Water: 23-28°C, pH value 6-8, 4-18°dGH.

Diet: Animal food, especially fish, is the preferred diet while earthworms and beef heart make a good substitute.

Spawning: Not known.

Features: This very large-growing pim is not suitable for the basic home aquarium. They are, however, quite comfortable in large show tanks in zoos, where they can live harmoniously with other catfish and tetras.

Unfortunately, the magnificent Goldy Pleco, Scobinancistrus aureatus, loses its attractive colours with age. But even older Goldy Plecos are impressive in appearance and really enrich large show tanks.
Photo: M.-P. & C. Piednoir

This picture shows a not-yet-fully-grown Goldy Pleco, Scobinancistrus aureatus, which has not changed its colours.
Photo: Dr. J Schmidt

Goldy Pleco

Species: *Scobinancistrus aureatus* BURGESS, 1994 (30cm).

Origin: Brazil, the Amazon.

Habitat: Flowing waters.

Aquarium: Large tanks from 2m upwards. It is a good idea to furnish the tank with bogwood, but not essential. Hiding places, which are so important for many catfish species, can therefore be created with clay pipes or pieces of slate. Good filtration and clear water are important.

Water: 26-29°C, pH value 5-7, 2-16°dGH.

Diet: A large proportion of its diet has to be animal food. Algae and other vegetable food are not of much interest to this species and it can survive without cellulose.

Spawning: Not known.

Features: This catfish is particularly colourful when young; with age the colour will turn black with fine white spots. The colour of the fins also matches the body's colour, which means the pretty red-orange tint will disappear completely. However, the colour of mature catfish is still attractive and is similar to that of a young *Acanthicus* or *Ancistrus*. Older Goldy Plecos become quite aggressive and will not tolerate any other members of its species kept in the same tank for company.

My Tip: If you don't have a large aquarium, stick to the smaller *Ancistrus* species, which are equally attractive but don't take up so much room.

The Duckbeak Shovelnosed Catfish, Sorubim lima, *has a very large mouth and an expandable stomach and can only be kept with larger fish like, for example, large cichlids.*
Photo:
Dr J. Schmidt

Duckbeak Shovelnosed Catfish

Species: *Sorubim lima* BLOCH & SCHNEIDER, 1806 (60cm).

Origin: The Amazon in Brazil, Paraguay and Venezuela.

Habitat: This robust predatory catfish is found in all types of water conditions, i.e. flowing as well as still clear, black and white waters.

Aquarium: Large tanks from 2m upwards are important; so is bogwood or long-leaved aquatic plants for cover, where this catfish likes to lie in wait for its prey or rest after it has eaten. Substrate type and grade are relatively unimportant.

Water: 24-28°C, pH value 6-7, 4-18°dGH.

Diet: Live and frozen food. Shovelnosed Catfish particularly like live fish, but will also eat earthworms, beef heart, food tablets and pellets.

Spawning: Not known.

Features: This predator, which does not swim actively, is only really suitable for the catfish expert. If, however, you have large breeding tanks, this fish can be quite useful to get rid of certain unwanted fish – such as stunted ones.

My Tip: Although this catfish has an unusual shape and therefore an attractive appearance, you are advised not to keep it in a home aquarium because of its size and feeding requirements.

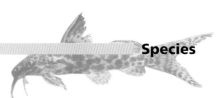

*Portrait of a Duckbeak Shovelnosed Catfish, Sorubim lima.
Photo:
Dr J. Schmidt*

*These catfish only feel really at home in very large aquariums like, for instance, show tanks in zoos.
Photo: M.-P. & C. Piednoir.*

The barbels of the upside-down catfish of the Synodontis genus are covered with papillae that help the fish to detect food in the water.
Photo: M.-P. & C. Piednoir

Black-Spotted Dusky Synodontis

Species: *Synodontis acanthomias* BOULENGER, 1899 (40cm).

Origin: The region of the Congo delta in West Africa.

Habitat: All water types.

Aquarium: Because of its size a larger aquarium of 150cm upwards is required. Bogwood and a dense plantation with large robust plants as hiding places are important, as these catfish become quite territorial with age and other fish can retreat from them. However, as young fish they are generally peaceable and enjoy swimming in a group.

Water: 23-27°C, pH value 6-8, 4-18°dGH.

Diet: Food tablets, left-overs from other fish foods and animal food like mosquito larvae, *Tubifex* and water fleas. Small fish are also considered as food.

Spawning: Not known.

Breeding: While breeding has not yet reportedly been successful, trials would be worthwhile, especially as these are popular fish in the aquatic hobby.

The Angel Catfish, Synodontis angelicus, is a very attractive catfish. Unfortunately, it is quite aggressive and quite rare in the wild.
Photo: Aqualife Taiwan

Angel Catfish

Species: *Synodontis angelicus* SCHILTHUIS, 1891 (18cm).

Origin: West Africa in the Congo and Cameroon.

Habitat: Clear- and black-water tributaries of large streams.

Aquarium: A single Angel Catfish can be kept in a 200-litre aquarium. As these catfish – despite contrary opinions in some of the literature – are quite cantankerous, a large aquarium from 2m upwards with plenty of hiding places is necessary.

Water: 23-28°C, pH value 6-8, 4-15° dGH.

Diet: Food tablets, left-overs from other fish foods and animal food, such as mosquito larvae or *Tubifex*.

Spawning: Not known.

Features: A sandy substrate is preferable, plants can then be set in pots.

My Tip: As this appears to be an endangered species, this catfish should only be kept if you are seriously interested in breeding the species.

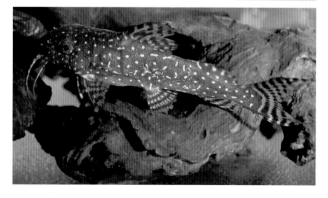

cichlids. When they eventually leave the cichlid's mouth, they are already more than 1cm long; they can be fed with normal food and are not in any danger from other fish in the tank.

Cuckoo Synodontis

Species: *Synodontis multipunctatus* BOULENGER, 1898 (12-15cm).

Origin: Lake Tanganyika in Africa.

Habitat: This upside-down catfish is actually rarely found in the lake but rather in the estuaries of the larger inflowing rivers (Malagarasi).

Aquarium: This catfish is best kept with Lake Tanganyika cichlids in large tanks with boulders and plenty of hiding places. A sandy substrate is favourable.

Water: 23-27°C, pH value 7-8.4, 15-35°dGH.

Diet: Food tablets, flakes, left-overs from other fish foods and animal food such as mosquito larvae, *Tubifex* worms or small shrimps.

Spawning: They spawn with mouthbrooding cichlids who take up the eggs with their own spawn.

Breeding: Breeding is usually successful when kept with mouthbrooding cichlids. While in the mouth of their foster mother, the two to five small catfish fry eat the tiny

My Tip: Because of its interesting and unusual breeding behaviour, this catfish will also appeal to cichlid enthusiasts. Although found in Lake Tanganyika in the wild, it is not essential to keep these catfish with Lake Tanganyika cichlids, breeding can also be successful with Lake Malawi cichlids or other mouthbrooders.

Trachelyichthys decaradiatus

Species: *Trachelyichthys decaradiatus* MEES, 1974 (12cm).

Origin: South America, in Guyana.

Habitat: Calm areas of flowing waters, especially with fallen wood and dense vegetation.

Aquarium: A 200-litre aquarium is adequate for the care of these catfish. They do not dig very much, but a sandy substrate is still advisable. Dense vegetation at the edges and hiding places made from stacked-up bogwood meet their need for shelter.

Water: 23-27°C, pH value 6-8, 4-15°dGH.

Diet: Food tablets, flakes, granulates, left-overs from other fish foods as well as types of live food.

Spawning: Not known.

My Tip: This species, which is most active during twilight or at night, can safely be kept with similarly sized, calm fish.

Pencil Catfish

Species: *Trichomycterus* sp. (6-15cm).

Origin: South America.

Habitat: All water types.

Aquarium: Tanks measuring from 1m upwards are suitable for a group of at least six catfish.

Water: 20-24°C, pH value 5-7, 4-18°dGH.

Diet: Food tablets, left-overs from other fish foods and animal food, such as mosquito larvae.

Spawning: Not known.

Features: The colour of these catfish is very variable, which makes identification without knowing their place of origin very difficult. There are some species in this family which are feared, as they tend to penetrate the urethras of any humans swimming in the water.

Trachelyichthys decaradiatus is suitable for a community aquarium. However, it is especially active at night.

The pencil catfish of the Trichomycteridae family are very difficult to define. Even the species name of these catfish, Trichomycterus sp., is unknown. Photos: H. Hieronimus

Concluding remarks

This book has introduced you to the most suitable catfish for aquatic purposes from the seemingly inexhaustible multitude of family groups. Although the main focus is on small species which are easy to keep in a home aquarium, I have consciously introduced you to other species which are often imported when young but soon outgrow the average home aquarium. Of course, it is never possible to cover all the important species in this type of book. It is, however, important that you should always bear in mind that you should never acquire a catfish that you know nothing about.

The very attractive catfish sold in many specialist shops are often the young of large-growing species. Even if you think you are missing a bargain, it is always better to get as much information as you can about any unfamiliar species. A good reputable aquatic shop which does not always know everything – just like every human being – will always recommend a good book in which to find all the information that you require.

Please remember: not even large zoo aquariums can accept unlimited numbers of fish, which have outgrown their tanks, from private owners.

After these words of caution I hope you will enjoy and be successful in your hobby. There is still plenty to discover, as new, completely unidentified species are regularly imported, and breeding has only been successful with few species, so that you too can dream of success in this fascinating area of aquatics.

Synodontis soloni belongs to a regular species but is very rarely imported.
Photo: Dr J. Schmidt

A Giant
Whiptail,
Sturisoma
aureum.
Photo:
Dr J. Schmidt.

BAENSCH, H. A. & RIEHL, R. *Aquarium Atlas Volumes 1-5* (Hans A. Baensch GmbH).

BARTHEM R. & GOULDING M. *The Catfish Connection: Ecology, Migration and Conservation of Amazon Predators* (Columbia University Press).

BURGESS, W. E. *An Atlas of Freshwater and Marine Catfishes: A Preliminary Survey of the Siluriformes* (T.F.H. Publications, Inc.).

BURGESS, W. E. *Colored Atlas of Miniature Catfish: Every Species of Corydoras – Brochis – Aspidoras* (T.F.H. Publications, Inc.).

BURGESS, W. E. *Corydoras Catfish* (T.F.H. Publications, Inc.).

EVERS, H.-G. & SEIDEL, I. *Wels Atlas Band 1* (in German)(Mergus GmbH).

FERRARIS, C. *Catfish in the Aquarium* (Tetra Press).

FULLER, I. *Breeding Corydoradine Catfishes* (Ian Fuller Enterprises).

GEIS, R. *Catfish: Keeping and Breeding Them in Captivity* (T.F.H. Publications, Inc.).

GLASER, U. *Aqualog Photo Collection 1 – African Catfishes* (Verlag A.C.S. GmbH).

GLASER, U. *Aqualog Special Loricariidae: The most beautiful L-Numbers* (Verlag A.C.S. GmbH).

GLASER, U. & GLASER, W. *Loricariidae: All L-Numbers* (Verlag A.C.S. GmbH).

GLASER, U., SCHÄFER, F. & GLASER, W. *All Corydoras* (Verlag A.C.S. GmbH).

GRANT, S. *Catfish Compendium* (A series of articles and pictures on a wide variety of catfish) (D.M.A. Wright).

JINKINGS, K. *Bristlenoses: Catfish with Character* (Kingdom Books).

KOBAYAGAWA, M. (ED. BURGESS, W. E.) *The World of Catfishes* (T.F.H. Publications, Inc.).

LAMBOURNE, D. *Corydoras Catfish* (Blandford Press).

SANDS, D. *Catfishes of the World, Vols. 1-5* (Dunure Publications).

SEUSS, W. 1993. *Corydoras – The Most Popular Armoured Catfishes of South America* (Dähne Verlag GmbH).

Enjoy all the great titles
in the AquaGuide series:

Aquarium Plants
ISBN 1-84286-034-8

Catfish
ISBN 1-84286-084-4

Discus
ISBN 1-84286-037-2

Freshwater Stingrays
ISBN 1-84286-083-6

Lake Malawi Cichlids
ISBN 1-84286-035-6

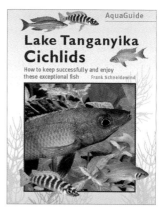

Lake Tanganyika Cichlids
ISBN 1-84286-036-4

For further information about these books and the full Interpet range
of aquatic and pet titles, please write to:

Interpet Publishing, Vincent Lane, Dorking, Surrey, RH4 3YX
email: publishing@interpet.co.uk Tel: +44 (0)1306 873822